Alternative Economic Spaces

Alternative Economic Spaces

Edited by
Andrew Leyshon, Roger Lee and Colin C. Williams

SAGE Publications
London • Thousand Oaks • New Delhi

First published 2003

SAGE Publications Ltd
6 Bonhill Street
London EC2A 4PU

SAGE Publications Inc.
2455 Teller Road
Thousand Oaks, California 91320

SAGE Publications India Pvt Ltd
B-42, Panchsheel Enclave
Post Box 4109
New Delhi 100 017

British Library Cataloguing in Publication data

A catalogue record for this book is available from the British Library

ISBN 0 7619 7128 9
ISBN 0 7619 7129 7 (pbk)

Library of Congress Control Number 2003102340

Typeset by C&M Digitals (P) Ltd, Chennai, India
Printed in Great Britain by Athenaeum Press, Gateshead

Contents

List of Contributors

Theresa Aldridge Department of Geography and Earth Sciences, Brunel University, Uxbridge UB8 3PH, UK

Ash Amin Department of Geography, University of Durham, South Road, Durham DH1 3LE, UK

Kate Brooks Department of Geography, University of Sheffield, Winter Street, Sheffield S10 2TN, UK

Angus Cameron Department of Geography, University of Leicester, University Road, Leicester LE1 7RH, UK

Louise Crewe School of Geography, University of Nottingham, University Park, Nottingham NG7 2RD, UK

Duncan Fuller Division of Geography and Environmental Management, Lipman Building, University of Northumbria, Newcastle-upon-Tyne NE1 8ST, UK

Nicky Gregson Department of Geography, University of Sheffield, Winter Street, Sheffield S10 2TN, UK

Ray Hudson Department of Geography, University of Durham, South Road, Durham DH1 3LE, UK

Jeffrey Jacob Graduate Division of Educational Research, The University of Calgary, 2500 University Drive, N.W., Calgary, Alberta T2N 1N4, Canada

Andrew E. G. Jonas Department of Geography, University of Hull, Cottingham Road, Hull HU6 7RX, UK

Roger Lee Department of Geography, Queen Mary, University of London, Mile End Road, London E1 4NS, UK

Andrew Leyshon School of Geography, University of Nottingham, University Park, Nottingham NG7 2RD, UK

Andrew Lincoln Department of Geography, Queen Mary, University of London, Mile End Road, London E1 4NS, UK

Jane Tooke Centre for Urban and Community Research, Goldsmiths College, University of London, New Cross, London SE14 6NW, UK

Colin C. Williams Department of Geography, University of Leicester, University Road, Leicester LE1 7RH, UK

Jan Windebank Department of French, University of Sheffield, Arts Tower, Western Bank, Sheffield S10 2TN, UK

Preface: Geography, diversity, change and ends

The geographies and histories of predicted ends remain, thankfully, to be written. What seems, however, to be true is that the prophets of recent predictions embody an apparently contradictory mix of a-geographic structural determinism and libertarian pretensions. What are conceived unproblematically as a-social (but liberal!) and in-variant markets have, at least in the views of the likes of Francis Fukuyama (1992), brought social change to an end as we now live in the best of all possible social worlds. Rather more realistically, others – equally impressed with the 'apparent triumph of the market' (Yergin and Stanislaw, 1998: 16) – point to certain social and material tests which must be met by 'the return to traditional liberalism around the world'.

Niall Ferguson (2000: 8) seems to go one social stage further in seeing 'modern economic history as a tale of capitalist triumph'. But that this socially-constructed link between capitalism and change is more apparent than real is revealed in his prior a-social acceptance of 'the fundamental notion that it is money – economics – that makes the world go round'. Indeed, he ridicules the notion that capitalism is a powerful social shaper of history and so falls back on an unproblematic separation of an unproblematic politics and economics. Further, although claiming that '[T]he nexus between economics and politics is the key to understanding the modern world' and that the 'causal connections between the economic and political world ... are so complex and so numerous that any attempt to reduce them to a model with reliable predictive power seems doomed to fail' (2002: 20), Ferguson's economic, political and social imagination is, like other similar luminaries, profoundly a-geographic and so fundamentally limited.

In contrast, the authors of this volume present rather less simplistic attempts to understand the world. This is not surprising. For one thing, they write from a geographically-sensitized perspective which recognizes the differentiated, multi-scalar and relational networks of change. For another, these essays present original research. Here, then, there is less a reliance on meta-generalization and notions of undifferentiated change than on detailed, theoretically-informed empirical research founded on the diverse social construction of economy and polity in which geographies are formative and generative of differentiated transformations.

The concern of the book is the presentation of research on a variety of ways in which people create and implement ways of practising economic life shaped and directed through sets of social relations differentiated from – and in some cases opposed to – mainstream relations. The research interest in these creative processes relates to the diversity of social relations underpinning economic activity, the degree to which, and the criteria by which, these practices may be judged to be 'successful', and the relations of influence between them and mainstream alternatives.

Above all, however, it is the celebration and analysis of the possibilities of diversity and economic proliferation that is the most significant contribution of the research reported here. Less a concern with closure and ends than with the perpetual opening out and transformation of social life. In the stultifying and unintelligent – to say nothing of the morally reprehensible – world of the assertion of hegemonic superiority, post-11 September 2001, these essays show how such diversity and proliferation are not only possible but reflect the irrepressible vitality and diversity of social life. And this despite the powerful pressures of conformity embedded in economic globalization.

The research reported here makes it clear that attempts to write geographies and histories of ends are, as yet, premature. And, in any case, if they are capable, ever, of being written, they must be able to take account of the micro-origins of social construction and resistance as well as of the multiple scales and meta-narratives of geography and history. It is the presentation of a number of attempts to flesh out such interpretations through empirical research that is the major purpose of this book.

References

Ferguson, N. (2000) *The Cash Nexus: Money and Power in the Modern World* 1700–2000. London: Penguin.

Fukuyama, F. (1992) *The End of History and the Last Man*. London: Penguin.

Yergin, D. and Stanislaw, I. (1998) *The Commanding Heights: The Battle between Government and the Market Place that is Remaking the Modern World*. New York: Simon & Schuster.

Roger Lee and Andrew Leyshon
London and Nottingham

Acknowledgements

The idea for this book emerged from discussions that surrounded an Economic and Social Research Council-funded project on Local Exchange Trading Systems (R: 000237208) that the editors helped to co-ordinate. We are very grateful for the contribution and support of Theresa Aldridge, Jane Tooke and Nigel Thrift, who were our fellow collaborators on that project, and who participated in numerous discussions about alternative economic spaces. Andrew Leyshon is also grateful to David Matless and graduate students in the School of Geography, University of Nottingham, for the provocative ideas about alternative economic geographies produced during discussion sessions in the Critical Human Geography and Current Research in Human Geography modules. We would also like to thank the contributors to this volume for their initial enthusiasm for participating in this project and for their patience in waiting for the final manuscript. Last, but certainly not least, we would like to express our thanks to Robert Rojek of Sage, who was enthusiastic and supportive at the outset and displayed a degree of forbearance and faith in the project that was well beyond the call of duty.

Acknowledgements

one

Introduction: Alternative Economic Geographies

Andrew Leyshon and Roger Lee

There was a time, during the 1990s, when the hegemony of a particularly virulent strand of global capitalism seemed assured and unquestioned. The strident assurances of advocates of neo-liberalization – perhaps the most notorious being that of a certain British Prime Minister who, a decade earlier, had stated that 'there is no alternative' to a mode of economic co-ordination based primarily upon market signals – seemed, in retrospect, to be less an example of dogmatic political rhetoric than a chilling prophesy of a monochrome global economic future. Economic globalization, the collapse into economic anarchy of state socialism in the Soviet Union and its satellites, and the emergence of a form of bureaucratic capitalism within states such as China and Vietnam, seemed to confirm the unfettered extension of capitalist mores and values into new spaces and places. Although even the harshest critic of capitalism should perhaps give pause to regret too strongly the passing of undemocratic and authoritarian social formations that prioritized the deferment of social consumption and rising living standards in favour of the seemingly perpetual reinvestment of surpluses within the means of production, the collapse of these systems had symbolic power because they also removed from view modes of economic and social reproduction that stood in contrast and in opposition to capitalism (see, for example, Bauman, 1992). It also meant the gradual erasure of economic geographies produced by non-capitalist social relations as these alternative economic spaces began to import market-based modes of co-ordination (Gowan, 1995).

At the same time, places that had long been characterized by capitalist social relations were witness to an intensification and increasing pervasiveness of such relations. Commentators on both left and right agreed that the core capitalist economies of the West had undergone something akin to a 'corporate takeover' of civil society (Crook, 2001a; Monbiot, 2000), as capitalism and its agents seemed to be determined to achieve what, to borrow from Habermas, could be described as the 'colonization of the lifeworld', by penetrating the nooks and crannies of everyday life (Hutton, 1995; Klein, 2000), which even included an attempt to colonize the precognitive domains of human subjects (Thrift, 2000).

Indeed, for some it was as if the disturbing predictions of the likes of Francis Fukuyama and Richard O'Brien were bearing fruit. Fukuyama (1989) argued that with the end of the Cold War a new era was beginning which represented, in Hegelian terms, the 'end of history', as societies abandoned 'their ideological pretensions of representing different and higher forms of human society' (1989: 13), and admitted the 'victory' of capitalism. As such, 'life for that part of the world that has reached the end of history is far more preoccupied with economics than with politics or strategy' (1989: 16), and involves societies around the world coalescing around a liberal, market-based model of social organization. Meanwhile, O'Brien (1991) claimed that a combination of information and communication technologies plus re-regulation was bringing about the end of geography within the global financial system, as it became increasingly straightforward to move money and information from place, and as the regulatory context within which exchange took place underwent convergence. For O'Brien, the 'End of Geography means [increased] competition' (1991: 115), and a concomitant speed-up in the pace of economic (and social) life (Reich, 2001; Schor, 1992).

On sober reflection, and in the wake of sustained and effective critiques, it was fair to say that the reports of the death of history (Derrida, 1994; Gray, 1998) and geography (Martin, 1994, 1999) were greatly exaggerated. Nevertheless, the arguments of Fukuyama and of O'Brien succeeded in striking a chord among many who witnessed with some unease the greater priority given to economic and financial matters in everyday life and the ways in which the spread of global corporations and products appeared to bring a new conformity, homogeneity and pace of life across capitalist societies. Moreover, this transition was actively supported by a capitalist elite that at every turn seemed willing to suggest that these transformations were not only in the best interest of society as a whole, but were also, incidentally, inevitable and unstoppable.

At the beginning of the twenty-first century, the celebrants and defenders of global capitalism are perhaps not as confident as they were. Cracks have begun to appear in the edifice of global capitalism. This statement might appear odd given that the capitalist system remains robust and durable, and actually underwent one of the longest periods of sustained growth in its history in the five years between 1995 and 2000, mainly on the back of a remarkable rise in the value of US financial assets (Brenner, 2000). Moreover, during this period the system was able to shrug off a sequence of major financial crises – including financial panics in Russia and the continuing crisis in Argentina and a more generalized financial crisis in East Asia (Kelly et al., 2001) – with the long bull market coming to an end only with the collapse in the share prices of 'technology' companies at the beginning of 2000 (Cassidy, 2002). Yet, it was actually during this period of long growth that a number of new theoretical and practical challenges to global capitalism arose to plant seeds of doubt in the minds of those who believed that the spread of neo-liberalized, global capitalism to all parts of the world was inevitable.

Indeed, David Harvey argues that opposition to capitalism flourished over this period to such an extent that it is now almost as widespread and ubiquitous as are capitalist social relations themselves. There is, he observes:

> ... not a region in the world where manifestations of anger and discontent with the capitalist system cannot be found. In some places or among some segments of a population, anti-capitalist movements are strongly implanted. Localized 'militant particularisms' ... are everywhere to be found, from the militia movements in the Michigan woods (much of it violently anti-capitalist and anti-corporate as well as racist and exclusionary) through the movements in countries such as Mexico, India and Brazil militating against World Bank development projects to the innumerable 'IMF riots' that have occurred throughout the world. (Harvey, 2000: 71)

Harvey argues that these protest movements should be seen as a response to what he describes as the 'utopian' post-war project to re-build international capitalism and, in particular, to the role played within this project since the 1950s by the 'Bretton Woods' institutions like the World Bank, the International Monetary Fund (IMF) (Harvey, 2000: 173–81), and more latterly the General Agreements on Tariffs and Trade (GATT) and its successor, the World Trade Organization (WTO). These institutions are key parts of what has become known as the 'Washington consensus', a programme of financial orthodoxy, privatization and deregulation which is seeking to reconstruct the world within a neo-liberal image (Peck and Tickell, 2002), the critics of which argue that it is restructuring the global economy in favour of rich economies at the direct expense of the poor (Fine, 1999). Harvey insists that such a project may be accurately described as utopian because it has a clear image of the future world that it wishes to create wherein things will be demonstrably 'better'. Indeed, it is this moral justification that underlies neo-liberal ideas that helps to explain why they have been adopted and implemented so readily, along with the fact that the 'programme' for the delivery of this utopia can be expressed in deceptively simple and straightforward terms. For its supporters, a market-based liberal economy is presented not merely as achievable but, because of the stress on its 'market' character and avoidance of its socially-constructed nature, as natural and, indeed, utopian. The quest for profit is made 'civilized' and becomes 'an engine of social progress' through the device of forcing 'firms ... to compete with their rivals for customers and workers' to value 'their reputation for quality and fair dealing – even if they do not value those things in themselves'. But even if they do not, '[c]ompetition will make them behave as if they did' (Crook, 2001b). Thus attempts to intervene in order to achieve particular objectives (for example, development through aid) are, in this view, not only undesirable but also doomed to failure. The liberal market is presented as a singular mechanism, the functioning of which can be affected only adversely by any attempt to intervene in it.

The critical point about such a representation of economic progress is that it avoids any notion of the social relations and relations of power though which all economic geographies take place and by which they are shaped. And yet, as is argued below, the social relations of economic geographies are inseparable from their material functioning and their material outcomes. Thus, the simplicity of the moral tale told above necessarily omits to point out that the road to this utopia is hard, and is littered with the casualties and losers in an often brutally competitive system. And it is the attempt to build this liberal, market-based utopia on a global scale that has, at least in part, sparked the recent wave of anti-capitalist and anti-globalization protests. Appropriately, then, given that the attempt to build global neo-liberalism is an inherently geographical project involving the construction of a uniform, neo-liberal global economic geography, the various oppositional movements and projects to 'think and perform the economy otherwise' reveal a keen attention to matters of space and place.

This edited collection draws attention to the efforts of individual and collective actors to imagine and, more importantly, to perform economic activities in a way that marks them out differently from the dictates and conventions of the mainstream economy. This introductory chapter provides a context for the book as a whole and is organized into two main parts. The next two sections consider the processes by which critics have sought to chip away at the edifice of neo-liberalism in order to challenge the primacy of this model and its prescriptive authority. This work has been of two main kinds, and we consider each in turn. First, there has been work of a primarily ideological and theoretical nature, which has sought to undermine the discursive foundation of neo-liberalism on its own terms. Secondly, there has been work of a more practical and performative nature, which has sought to demonstrate ways of enacting the economy otherwise. The division between these two oppositional strategies is, in reality, somewhat artificial, as they draw strength and inspiration from each other and indeed, if anything, the flow of ideas is often stronger from the latter to the former than the other way around. The final section of the introduction presents a synopsis of the book's structure and a brief discussion of each chapter.

THINKING THE ECONOMY OTHERWISE

Theoretical attempts to critique and undermine capitalist social relation are, of course, hardly new. Marxism and political economy have a long and honourable tradition in this regard, although their authority as a critical project has been undermined in recent years, both because of the collapse of actually-existing socialism and, more significantly, because of the rise of alternative critical traditions within social theory that have abandoned many of Marxism's certainties and its central concern with class in favour of a more relativistic, more discursive and generally contextual approach to enquiry. Taken together, these tendencies are usually described as the 'cultural turn'. The cultural turn's demotion of the category of class to a level that is no longer privileged above other social markers such as ethnicity, gender or sexuality has been criticized for bringing about a relegation of economic matters within critical debates at a time of deepening uneven development and income equality across a broad range of geographical scales (Hamnett, 2000; Ray and Sayer, 1999; Thrift, 1992). However, in contrast, we wish to argue that far from being a dilettante distraction from 'serious' political economy that many on the left fear, the cultural turn has been responsible for opening up new and effective forms of critique that have produced new strategies for undermining the power and authority of capitalism and its agents. There are numerous examples upon which we could draw, but we shall limit our illustrations to just two particularly effective examples of this kind of work: first, Gibson-Graham's work on the proliferative economy and, secondly, Carrier and Miller's work on 'virtualism'.

The proliferative economy

J.K. Gibson-Graham – a hybrid subject made up of Kathie Gibson and Julie Graham – draws upon the radical possibilities of the cultural turn to imagine alternative conceptions to a hegemonic and all-conquering 'global capitalist economy' (Gibson-Graham, 1996). That is, she uses cultural theory to argue that what is taken for granted within economic analysis is not quite so obvious when one takes inspiration from other critiques of dominant discourses. Inspired by the theoretical incursions made by feminist, racial and queer theory, it began to dawn on Gibson-Graham that

> a theoretical option now presented itself, one that could make a (revolutionary) difference: to depict economic discourse as hegemonized while rendering the social world as economically differentiated and complex. It was possible, I realized, and potentially productive to understand capitalist hegemony as a (dominant) discourse rather than as social articulation or structure. Thus one might represent economic practice as comprising a rich diversity of capitalist and non-capitalist activities and argue that the non-capitalist ones had been relatively 'invisible' because the concepts and discourses that could make them 'visible' have themselves been marginalized and suppressed. (1996: x–xi)

She argues that it is critically effective to think beyond the 'capitalocentricism' that has characterized conventional political economy writing in order to raise the possibility of dwelling within 'a noncapitalist place'. There is, she argues, within conventional approaches something approaching a double-deceit, wherein the economy is depicted as something that we can understand but over which we ultimately have no control. Therefore, Gibson-Graham seeks to build an alternative discourse of the economy, one that requires that we step outside its conventional bounds and begin to associate with new modes of thought and critique.

One particularly fruitful course suggested by Gibson-Graham is for economic theorists to learn from feminism and studies of sexuality in order to 'go beyond' the boundedness of their subject and to become both more recursive and reflexive, and to problematize the idea of 'the economic':

> Whereas feminist theorists have scrutinized and often dispensed with the understanding of the body as a bounded and hierarchically structured totality, most speakers of 'economics' do not problematize the nature of the discursive entity with which they are engaged. Instead, they tend to appropriate unproblematically an object of knowledge and to be constructed thereby as its discursive subjects. In familiar but paradoxical ways, their subjectivity is constituted by the economy which is their object: they must obey it, yet it is subject to their control; they can fully understand it and, indeed, capture its dynamics in theories and models, yet they may adjust it only in minimal ways. These experiential constants of 'the economy' delineate our subjective relation to its familiar and unproblematic being. (Gibson-Graham, 1996: 96)

Therefore, mounting an effective critique of capitalism, and imagining alternative economic spaces, requires a radical rethinking of the 'the economic object' (Gibson-Graham, 1996: 97). In particular, Gibson-Graham advocates a deployment of Foucault's genealogical method within economic analysis, so that it is possible to trace the evolution and development of the discursive formations that support and sustain the contemporary capitalist economy. She draws attention to the ways in which, over a long period of time, economic discourse both became strongly associated with the masculine and successfully imported a number of biological and physiological metaphors to sustain it. This has had at least two main implications. First, the historical tendencies to assume gendered mind/body dualisms, wherein the masculine is associated with the mind and reason and the feminine with passion and nature, etc. (Massey, 1995), ensured that the economy became associated with reason, rationality and order, to which 'the irrational disorder of non-economic life must submit' (Gibson-Graham, 1996: 103). Secondly, the central role of biological and physiological metaphors within economic discourse encouraged the capitalist economy to be thought of as an integrated and total ecology, one that effectively 'crowds out' other forms of economic reproduction. Moreover, the central role of naturalistic metaphors within economic discourse has produced a mode of thought that draws upon traditional evolutionary theory to suggest that economic forms evolve in a linear manner by passing through different stages to end up at a more 'advanced' or 'developed' stage.

Gibson-Graham seeks to expose the metaphors and assumptions that underpin conventional economic discourse, so as to pick them apart, reveal their contingent and contextual qualities and, thereby, undermine their power. What she seeks to do, above all else, is to disrupt and destabilize the irrevocable association of 'economy' with 'capitalism' in order to produce a mode of thinking that considers the economy to be 'proliferative rather than reductive of forms' (Gibson-Graham, 1996: 119). She evokes the ideas of contemporary evolutionary theorist Steven Jay Gould to argue that economic and social development does not proceed in a linear or successive fashion but, again, is proliferative. This observation has important political implications for imaginings of alternative economic systems, and provides crucial intellectual resources for envisioning an economic world wherein there are, indeed, alternatives.

However, as soon as the inseparable and mutually formative relations between the material and the social in the construction and functioning of economic geographies are recognized (for example, Lee, 1989, 2002), this is not an altogether surprising conclusion. Economic geographies are circuits of consumption, exchange and production sustained over space and through time. The purely material possibilities for proliferation are immediately apparent from such a statement. Yet, the neo-liberal agenda argues for a market-induced economic-geographical uniformity and singularity. The assumption here is of an a-social economy controlled entirely by unfettered market exchanges. This is ironic as the particular social construction of the economic advocated by the neo-liberal agenda is itself a social construct – normally unproblematized – that seeks to deny any alternative. Such a denial is actively achieved both through the persistent representation of economic geographies merely as market mechanisms and through the practices of power wielded at a global scale over such geographies by institutions such as the World Bank, the IMF and the WTO in the attempt to ensure that that is what they remain.

However, all economies and economic geographies are both material *and* social constructs. Thus the possibilities for proliferation themselves proliferate well beyond the purely material. Economies based upon the particularities of capitalism (involving a particular set of class relations and the objective of accumulation) may exist alongside those based, for example, upon mutuality, ecological sustainability and social justice. However, all economic geographies must (always) be constrained by the requirements of materially effective circuits of consumption, exchange and production. This is simply because if economies are not capable of consuming, exchanging and producing use values, they are incapable of sustaining the means of social reproduction and so are doomed to fail. But to say this is not to say that effective economic geographies are reducible to a singular model. The scope for proliferation, though constrained, is still very wide.

This argument is strongly supported by James Ferguson (Ferguson, 1999) in his analysis of the fate of the Zambian copper belt following the collapse of the price of copper in world markets and the subsequent decrepitude of its mining industry and the growing burdens of external debt and IMF intervention. These developments have condemned the Zambian economy as a whole to rapid and wholesale economic decline. Amidst this crisis of development, and the disappointment, deprivation and despair so engendered, Ferguson draws attention to the ways in which, as a matter of survival, individuals and households switch to seemingly forgotten and redundant modes of economic and social behaviour. In other words, older modes of economic and social organization are revealed never truly to die out, as conventional evolutionary accounts would suggest, but to merely retreat into the background as relic forms that are overshadowed by more 'modern' modes of development. However, these older ideas and practices may resurface in times of economic and social crisis, with the result that 'the 'dead ends' of the past keep coming back, just as the 'main lines' that are supposed to lead to the future continually seem to disappoint' (Ferguson, 1999: 251). Thus, to survive, inhabitants of the Zambian copper belt have revived forms of economic behaviour – such as circular migration and shifting cultivation – thought to have died out when Zambia, and Africa more generally, embarked upon its path towards development. As Ferguson makes clear, these so-called 'dead-ends' represent vital resources and strategies for the survival of individuals and households experiencing the 'abjection' of disconnection from a global economy to which they were once gainfully attached. For Ferguson, the political and theoretical implications of such developments are clear:

> ... challenging neo-liberal globalization cannot simply be a matter of confronting it with its successor (the next historical stage, a higher rung on the ladder) but must involve working through the 'full house,' the actually existing 'bush,' of partly overlapping social forces and organized movements that are at work on different visions of the 'new world order.' New social movements mobilized around such issues as ecology, sexuality, religion, and human rights can take their place here alongside revitalized Marxist critique, a re-energized global labor movement, a politicized humanitarianism, even a rejuvenated Keynesianism. Emerging new forms of resistance to the brutalities of global capitalism, that is, must coexist with older forms, scrounged ... from the dustbin of history. (Ferguson, 1999: 257)

Work of this kind therefore seeks to undermine the privileged status of neo-liberal discourse and to open up a space to both discuss and practise non-capitalist modes of economic reproduction.

It is complemented by a second strand of work that also seeks to destabilize the discursive formation of neo-liberalization. A particularly good example is the development of the concept of 'virtualism by the anthropologists James Carrier and Danny Miller, to which we now turn our attention.

Virtualism

Carrier and Miller set out to illustrate the power of abstract theory within the material reality of economic social life and to argue that, through the process of neo-liberalization, the economic realm is becoming more abstracted from its pre-existing social and political contexts (Carrier and Miller, 1998; Carrier, 1998; Miller, 1998, 2000). Virtualism is defined as 'the conscious attempt to make the real world conform to the virtual image, justified by the claim that the failure of the real to conform to the ideal is a consequence not merely of imperfections, but is a failure that itself has undesirable consequences' (Carrier, 1998: 8). Reinforcing Harvey's insistence that neo-liberalism is driven by an utopian impulse and the drive towards abstraction (2000: 241–2), Carrier argues that economic thought and practice influence one another in an ongoing recursive and reflexive loop, which is 'driven by ideas and idealism [and] the desire to make the world conform to the image' (Carrier, 1998: 5). These abstractions have become more powerful because of the way in which they are supported by the academic discipline of economics, which has successfully colonized a number of key economic and political institutions so that they assist with the propagation of such abstractions.

For Miller, the power of abstraction contained within virtualism is an alienating force, and one that is bringing about a mutation within capitalism, which has become a more regressive and polarized social formation than it was during much of the post-war period of the twentieth century. Thus, the quest for an abstracted ideal of a market-economy pursued through the political project of neo-liberalism has eroded many of the social gains made from the 1930s onwards when governments sought to temper the impacts of markets through a series of social compromises and settlements (Polanyi, 1957). The creation of the welfare state during this period did much to negate many of the alienating impacts of capitalism, as levels of income rose and growing sections of the population began to enjoy increases in disposable income and access to a greater volume of consumer goods. Subverting the idea within Regulation Theory that mass consumption was necessary to 'balance' the forces of mass production developed in the earlier part of the twentieth century, Miller argues that mass consumption enabled capitalism to enter a period of relative stability within the post-war period not just because it provided a market for the goods and services so produced, but more because the benefits of participation within mass consumption – supported by the intervention of the welfare state in many economies – by the majority of the population fended off the alienating tendencies of capitalism (Miller, 1998: 193).

However, through the 1980s and 1990s neo-liberalism unleashed the effects of virtualism, bringing about new extremes of income inequality on a global scale. Examples of virtualism include models of structural adjustment, inspired by neo-classical economics, and the rise of a generalized audit society. What Miller reveals is that although these projects have had significantly alienating effects, they have, as indicated above, also been introduced with a moral agenda of making social gains. The power of neo-liberal economic theory within virtualism has ensured that these gains are validated at the level of the individual, constructed as a consumer. But, as Miller points out, these gains are claimed not in the name of actual, living consumers but, rather, in the name of abstracted, idealized or virtual consumers:

> Since it was consumption as an expression of welfare that was the main instrument in negating the abstraction of capitalism, the move to greater abstraction had to supplant consumption as human practice with an abstract version of the consumer. The result is the creation of the virtual consumer in economic theory, a chimera, the constituent parts of which are utterly daft. ... Indeed, neo-classical economists make no claim to represent flesh-and-blood consumers. They claim that their consumers are merely aggregate figures used in modelling. Their protestations of innocence are hollow, however, because these virtual consumers and the models they inhabit and that animate them are the same models that are used to justify forcing actual consumers to behave like their virtual counterparts. Just as the problem with structural adjustment is not that it is based on academic theory but that it has become practice, so the problem with the neo-classical consumer is the effects that the model has on the possibilities of consumer practice. In some kind of global card trick, an abstract, virtual consumer steals the authority that had been accumulated for workers in their other role as consumers. (Miller, 1998: 200)

In other words, in the name of lower prices, greater 'efficiency' and a greater *freedom to* make economic choices by those who can afford it, economic virtualism has removed many of the substantive social advances and improvements brought about by the welfare state and reduced the number of people that enjoyed *freedom from* poverty and deprivation.

Miller insists that the problem is not so much the creation of abstractions as such, but the way in which abstractions are afforded such power within the practical projects that politicians and administrators execute in our name. There is, therefore, a clear need for intellectual work that directly confronts such abstractions, and prises them open to scrutiny so that they 'can be re-appropriated as part of the enhancement of human self-understanding and cultural development' (1998: 212).

There are numerous other examples of culturally inflected economic research that seek to open up a space for thinking creatively about opposing neo-liberal capitalism and imagined alternatives. One particularly germane example is the work of Actor Network Theory (ANT), given both its stated intention to open up the 'black boxes' created by the unfolding of socio-technical networks to make possible a world configured otherwise (see Bingham and Thrift, 2000), as well as its growing interest in the economic sphere (Callon, 1998).[1] Indeed, this conception of economic activities as networked achievements that are spread across space encourages a mode of thinking that is strongly reminiscent of the Gibson-Graham project, which is to encourage a way of seeing capitalism as a more fragile and vulnerable entity than hitherto. Thus, imagining capitalism as a network in the manner of ANT is contrary to the conventional 'understanding of capitalism as a unitary figure coextensive with the geographical space of the nation state (if not the world)' and helps develop a geographical imaginary wherein capitalism is depicted as a 'desegregated and diverse set of practices unevenly distributed across a varied economic landscape' (Gibson-Graham, 1996: 117). In other words, by re-imagining capitalism as a network that has constantly to be achieved, it becomes possible to identify those places within space economies where the network is very weak, and where potential exists for new forms of alliances, social formations and economic geographies first to take root, then to become established, and finally to flower and bloom.

Constraints of space mean that we are unable to develop these ideas in any detail here, although some practical examples of this process of working away at such points of weakness and least resistance are developed in the next section of the chapter. Before we move on to consider these examples of 'enacting the economy otherwise', it is also important to note that the cultural turn has, alongside the collapse of state socialism, been responsible in large part for an internal re-assessment and re-statement of the Marxist project itself. In doing so, it has also engendered a renewed interest in the possibilities of alternatives to capitalism that go beyond the failed models of state socialism. For example, David Harvey points out that while a large number of critics have drawn attention to the failings and problems associated with a life dominated by capitalist social relations, hardly any of these critics have been prepared to say what kind of alternative world they would like to see inserted in capitalism's stead (Harvey, 2000). The reluctance of these critics to do so, Harvey insists, stems from the way the cultural turn has so privileged a sensitivity to difference that it has disabled the possibility of suggesting alternative worlds, lest some aspect of them be accused of being oppressive or repressive of various kinds of cultural and political expression. This hesitation reveals for Harvey an inability to arbitrate, which has unfortunate political implications given the advance of neo-liberal projects that are unhindered by an openness to alternatives:

> The anti-authoritarianism of liberatory political thought ... reaches some sort of limit. There is a failure to recognize that materialization of anything requires, at least for a time, closure around a particular set of institutional arrangements and a particular spatial form and that the act of closure is itself a material statement that carries its own authority in human affairs. What the abandonment of all talk of Utopia on the left has done is leave the question of valid and legitimate authority in abeyance (or, more exactly, to leave it to the moralisms of the conservatives – both of the neoliberal and religious variety). It has left the concept of Utopia ... as a pure signifier without any meaningful referent in the material world. And for many contemporary theorists ... that is where the concept can and should remain: as a pure signifier of hope destined never to acquire a material referent. But the problem is that without a vision of Utopia there is no way to define that port to which we might want to sail. (Harvey, 2000: 188–9)

Thus, Harvey argues that the left has abandoned any claim to the idea of utopia, due to what he sees as a fear of engaging in an act of political arbitration that would prioritize some values over others. But not do so, he argues, sustains the deceit of living in what he describes as a 'both/and' world; ultimately, he insists, and as the protagonists of neo-liberalism clearly recognize, we live in an 'either/or' world where certain hard choices need to be made. This is because, as argued above, economic geographies are simultaneously – and inseparably – sets of social and material processes and relations. Critics of capitalism therefore need to set out practical visions of a non-capitalist world, if only as a mechanism of encouraging the idea that there can indeed be such an alternative. This position is strongly supported by other critics, such as Andrew Sayer for example (Sayer, 1995), who argues that Marxists have traditionally resisted setting out what their vision of a post-capitalist world might look like, as they wished to avoid prejudging the course that class struggle might take. As Sayer points out, this refusal to project into the future is curious, arguing that it is 'strange that a critical social science which is so exacting in its explanations and critiques of what is, should be so lacking regarding what could or should be' (Sayer, 1995: 35). However, as his careful critique of different modes of economic co-ordination reveals, there are far more possibilities available than the conventional divide between 'markets' and 'planning' and that the potential to develop 'possibilities' are 'less constrained than is commonly assumed on the Left' (Sayer, 1995: 182).

Both Harvey and Sayer draw particular attention to the dialectical tension that exists between processes of economic abstraction on the one hand and the more substantive processes of performing economies on the other. They argue that such a tension exists both within the formal or mainstream economy and within the struggles to create alternative economies of different kinds. In addition, they both point to the fact that these tensions, and the constructions and performances of mainstream and alternative economies, are necessarily played out in economic space. Indeed, Harvey argues that utopian ideas of an alternative to capitalism are highly geographical, often associated with an attempt to relocate 'beyond' the heartlands of capitalism, which are seen to be understandable – although, Harvey believes, ultimately futile – efforts to defend such experiments against the social relations of the outside world. The next section of this chapter looks at some examples of these substantive attempts to create alternative economic worlds.

PERFORMING THE ECONOMY OTHERWISE

There are numerous examples of practical attempts to challenge the hegemony of neo-liberal capitalism that begin to explore the possibility of organizing economic life otherwise. Perhaps the most spectacular has been the emerging consequences of the anti-capitalist protests that have taken place since the end of the 1990s. These protests move around the world, both shadowing and drawing attention to the ways in which global economic and political elites meet at regular intervals to co-ordinate the governance of the global economy. The third ministerial meeting of the World Trade Organization in Seattle in December 1999 marked the beginning of an extraordinarily visible wave of protest against the effects of neo-liberalism and capitalist globalization. The meeting was a failure on its own terms, as it collapsed without agreement on a new round of trade protocols.

However, this was rather overshadowed by the events outside the conference where over 100,000 people marched in peaceful protest to the meeting. The extensive media coverage of the event was promoted by the fact that the demonstrators were met by police in riot gear, leading to three days of pitched battles between the police and a minority of protestors, which produced over 500 arrests, a state of emergency within Seattle, and around $3 million of damage to property in the city. The outlets of particularly ubiquitous capitalist corporations, such as McDonalds and Gap, were singled out for particular attention in this regard (Fannin et al., 2000; Wainwright et al., 2000). However, of more significance is the fact that the Seattle demonstration was so well attended, and that it set in motion a wave of protests that disturbed and unsettled subsequent meetings of this nature across the world (Smith, 2000). Thus, over the next 18 months, the meetings of the World Bank and International Monetary Fund in Washington DC and in Prague, the World Economic Forum's annual meeting in Davos, the Summit of the Americas in Quebec City, a European Union expansion meeting in Gothenburg, and the annual meeting of the G8 in Genoa were all subject to vociferous – and at times violent – protests.

The scale of these protests, and their signification of a broader dissatisfaction with the progress of global neo-liberalism, has at least given leading politicians and capitalist representatives pause for thought. Reactions range from a blanket condemnation of the protests by leaders such as British Prime Minister Tony Blair and a vigorous defence of the moral authority of globalization by *The Economist* magazine (Crook, 2001b) to a willingness by the then German Chancellor Gerhard Schroeder and French Prime Minister Lionel Jospin to countenance extra regulatory controls on the worst excesses of 'fast capitalism', such as the imposition of a tax on currency speculation (Elliot, 2001; Hooper, 2001).

These anti-capitalist and anti-globalization protests also draw attention to the importance of place to the co-ordination of global capitalism and its successful governance, and the fact that it requires the co-presence of global political and economic leaders to thrash out conventions and agreements for neo-liberal capitalism to continue to go on. As the increasingly well-organized demonstrators clearly recognize, this temporary 'spatial fix' is a point of weakness within the network of global capitalism. However, this is less because capital has shown a historical tendency to distance itself from challenges to its authority – from the labour movement, for example – through the simple expedient of reorganizing its assets over space (Harvey, 1982) than because a multi-local capitalism is difficult to regulate. Regulatory economic summits are therefore continuously necessary; they have to take place and, for the regulators, there is nowhere to hide. No doubt, as the organizers of such meetings obtain more experience of dealing with and subduing these protests, the ability of protestors actually to disrupt the course of the meetings will decline. Yet, as long as the protests continue so they will serve to remind people that the progress of global neo-liberalism should not be taken for granted and that, ultimately, its ability to survive requires that it continues to achieve a degree of political legitimacy, at least within the core capitalist countries of the West.

However, in addition to the attempt to curb such protest from without, the anti-capitalist, anti-globalization movement also suffers from the problem that Harvey argues infects most critical studies of capitalism: the protestors are more united by what they are against than what they are for. As Crook (2001b) has observed: 'The main things holding the anti-globalist coalition together are a suspicion of markets, a strongly collectivist instinct and a belief in protest as a form of moral uplift'. However, he argues that the 'protest coalition can hang together only if it continues to avoid thinking about what it might be in favour of'.

More work in this direction is clearly needed, although if inspiration is required perhaps it can be gleaned from the numerous examples of practical, day-to-day experiments in performing the economy otherwise. Such opposition is demonstrated in various ways, such as participation in 'elective' alternative social movements and a modification of lifestyle, participation in consumer boycotts, or using greater discretion in acts of consumption (Hartwick, 2000). Such acts are typically not directly confrontational – although there is anecdotal evidence that many of the key movers within anti-capitalist protest are members of new social movements – but aim to bring about change in the nature of the economy through everyday practices. Thus, these actions are seeking to influence the day-to-day rhythms and cycles of the capitalist economy, and by so doing, divert and reshape them. Perhaps the best example of these activities is the fair trade movement that has developed from critical analysis of the production chains of commodities such as coffee, tea and clothing (see Klein, 2000). Building upon academic work on the circulation of commodities through value chains (Cook and Crang, 1996; Dicken and Hassler, 2000; Dicken et al., 2001; Gereffi and Korzeniewicz, 1994; Hartwick, 1998, 2000; Hughes, 2000; Jackson, 1999; Mather, 1999; Robbins, 1999), organizations such as Fair Trade and Clean Clothes

(Ross, 1997) attempt to make visible the passage of commodities from (developing country) producer to (western) consumer, and of the social relations of exploitation that accompany them. These organizations seek to replace these exploitative relations with 'fairer' terms of exchange, which involve producers obtaining better terms of trade than if they sold their products directly to large global corporations. Such movements have a clear educative purpose, and exist in the belief that consumers in the West would purchase in a more ethically responsible manner if only they were aware of the ways in which their purchasing decisions feed through these commodity chains to support and sustain exploitation and impoverishment elsewhere. Indeed, these educative aims are currently being taken further in Britain where, according to Thrift, efforts are in process

> to get the subject of commodity chains introduced into schools at Key Stage 3 of the National Curriculum, with the idea of using permanently-sited webcams which would allow schoolchildren to see each part of the commodity chain, thereby not only registering the geography of the commodity but also making them think about their own role as responsible consumers since they are able to see exactly what labour they are in part responsible for. (Thrift, 2002: 5)

At the same time, however, financial and entrepreneurial education is also advocated within the school curriculum.

Akin to such developments is the emergence of 'green' and 'ethical' development, which allow investors to profit from the proceeds of capitalism but only in ways consistent with their desire to avoid investing in the stocks and shares of companies considered to transgress various standards of 'appropriate' behaviour (Lewis and Mackenzie, 2000a, 2000b; Winnett and Lewis, 2000). Clearly, these examples are merely indicative and illustrative. There exist a range of other 'everyday' examples of performing and practising the economy otherwise, and they are the subject of the chapters of this book. Thus, the final section of the introduction provides a synopsis of their content.

ALTERNATIVE ECONOMIC SPACES

One of the key observations to emerge from the contents of this book is that the idea and concept of the economic alternative is highly unstable and relational. This is made clear in Chapter 4 by Louise Crewe, Nicky Gregson and Kate Brooks, which analyses the retailing of second-hand, 'retro' clothing. As they argue, the notion of the 'alternative' is a chaotic conception, and one that varies over space, through time and by industrial sector. Thus, many of those who work within 'creative' or cultural industries perceive what they do to be alternative or oppositional to 'mainstream' economic activities, although by the measure of some of the other practices described in this book, their activities would appear to be strongly implicated within capitalist social relations. The self-understanding of the actions performed by the respondents in Crewe, Gregson and Brooks's study as 'alternative' is based upon the importance placed upon creativity, autonomy at work and a general anti-corporate attitude. Thus, retro retailers, which are typically small-scale operations, seek to carve out an economic living for themselves through the deployment of tacit and situated knowledge by revalorizing and re-commodifying clothing; or, in other words, by making unfashionable clothes fashionable again. In doing so, they necessarily operate – initially at least – in opposition to the mainstream retail sector by offering clothing that is not normally available elsewhere. However, to the extent that such retro retailers are successful in revalorizing certain fashions, they are constantly in danger of seeing their market being colonized by larger, more economically powerful outlets. Moreover, mainstream organizations and agents often move not only into the 'retro' market but also into the geographical areas in which they operate, which have often been successfully revalorized as new, alternative economic spaces. This makes retro retailing an often precarious means of livelihood, although many of the respondents in the study had made the conscious decision to opt for a more difficult but more autonomous life, within which they would have greater freedom at work, rather than submit to what they saw as the demands and dictates of large, impersonal, bureaucratic organizations. Therefore, the alternative economic strategies pursued by such agents cannot be seen as oppositional to capitalism in general; after all, most of the respondents in the study could accurately be described as members of a proto-petty bourgeoisie. Rather, the efforts of the retro retailers, and other similar creative workers, should be seen as a means of economic reproduction that is developed as an alternative to the hegemony of corporate capitalism and to the powers that such institutions exert over their employees and which, for these respondents at least, made the social relations of employment unacceptable.

Chapter 5, by Andrew Lincoln, looks at institutions that have sought to transform the social relations of employment through the device of employee-ownership. Over the last 20 years or so, there has been an increase in the number of worker-owned enterprises within economies such as that of the United Kingdom, which Lincoln argues is a product not so much of efforts to create more democratic workplaces than of defensive strategies developed in response to programmes of privatization within state-owned industries. Worker ownership, and the development of institutional forms like co-operatives, has a long history, dating back at least to the eighteenth century, when it developed as a response to the emergence of an impoverished working class that arose in the wake of the Industrial Revolution. Worker ownership gives employees an opportunity of intervening within economic globalization, and of exerting some degree of control and influence over local economic processes. In other words, worker ownership holds out the possibility of exercising some agency within the structures of global capitalism. By securing some degree of local control and a stake within economic organizations, worker ownership has the potential to fix capital within a locality and, through mechanisms such as preferential purchasing decisions, perhaps to help ensure that capital circulates deeper and faster within a local economy than might otherwise be the case. However, while the examples of worker-owned organizations outlined by Lincoln demonstrate that it is possible to create alternative economic worlds, it is far more difficult to sustain them. Moreover, such worlds can be better sustained in some places than others and their survival, even within accommodating local milieux, requires its subjects to work out a means of negotiating an economic environment that establishes rules and norms of competitive behaviour that are often inimical to the more collective and democratic modes of governance that characterize worker-owned enterprises.

Chapter 2, by Ash Amin, Angus Cameron and Ray Hudson, undertakes a critical analysis of the social economy and, in particular, of the ways in which it has increasingly become a motor of local economic regeneration. The social economy has been projected as a means of moving between the market and the state, by drawing upon resources within local communities. Thus, the social economy is increasingly considered to be an effective modulator of the extremes of market capitalism and facilitates the development of initiatives that present alternative economic options for local communities, which is the level at which many initiatives are scaled. However, as Amin, Cameron and Hudson point out, the social economy is highly dependent upon the state and its agents, which provide most of its funding and much of its ideological support. For this reason, the idea of the social economy as a kind of radical economic alternative is deeply suspect. Indeed, one of the reasons that the state is so interested in the social economy is that it is a vehicle by which social and economic risk can be moved away from the state and on to local communities, which are expected to assume responsibility for the operation of social economy initiatives, often resulting in the self-exploitation of those involved through the allocation of considerable amounts of time and effort and the foregoing of other potential forms of income. Nor is there much evidence that these schemes deliver the benefits to local communities that many activists claim for them. According to Amin, Cameron and Hudson, the main beneficiaries of the social economy are the state (in its ability to abrogate responsibility for local economic development), the private sector (which may benefit from social economy initiatives), and an emerging cadre of social economy professionals. Thus, rather than being a means of developing radically alternative economic spaces, Amin, Cameron and Hudson argue, after Marx and Engels, that the social economy is perhaps better seen as a form of 'conservative' or 'bourgeois' socialism that seeks to moderate the effects of capitalism without actually reforming or replacing it.

A further account of alternative economic institutions emerges from Andrew Jonas and Duncan Fuller's analysis of credit unions (Chapter 3). Their chapter examines the ways in which alternative social forms are produced, and argues that they need to be understood relationally, being located within wider economic and political developments. As such, they propose a more nuanced understanding of alternative institutions, along a spectrum of 'oppositionality' to the norms of the mainstream economy. Thus, alternative institutions may be categorized as 'alternative-oppositional', 'alternative-additional' or 'alternative-substitute'. Credit unions, which are financial institutions that provide cheap credit to individuals and households on low incomes, have traditionally been examples of 'alternative-oppositional' institutions. They provide access to cheap credit, which would otherwise be denied to such individuals and households, with beneficial social and economic outcomes. However, as they argue the relative success of credit unions has led to them being drawn into the kind of policy debates that surround the social economy in general, so that a struggle is emerging between different models of credit union. On the one hand, there is an 'idealist' model that advocates local autonomy and empowerment, and on the other an 'instrumentalist' model that emphasizes economic viability. The struggle seems to be being won by advocates of the 'instrumentalist' model, resulting in a rescaling of the credit union movement, which is moving away from a more overtly local, community focus towards the building of larger, more 'efficient' institutions. As Jonas and Fuller argue, for many credit union activists this rescaling of the movement represents a betrayal of its former community-based, socially-oriented *raison d'être*, and so to a downgrading of the degree to which credit unions can create and sustain alternative economic space.

The conception of the economic alternative is broadened in Colin Williams and Jan Windebank's chapter on informal employment (Chapter 6). For Williams and Windebank, alternative economic space is all that which lies beyond formal employment; that is, the non-market production, consumption or exchange of goods and services. By focusing upon the large amount of work that goes on beyond formal employment, they seek to decentre the mainstream from its core position within conceptual understandings of the economy. In their chapter they set out to understand the uneven geographies of work that lie beyond formal employment, which includes paid informal work (which is hidden from the state for tax or other reasons), self-provisioning (unpaid work undertaken for one's own household) and mutual aid (unpaid work undertaken for other than one's own household). Williams and Windebank take inspiration from Amartya Sen to argue that social inequality is best analysed in terms of the 'capabilities' of households to cope and to reproduce themselves, which is not necessarily dependent upon conventional measures of economic well-being such as income. Thus, some households with relatively low levels of income may be able to demonstrate considerable 'capabilities' due to their ability to draw upon broad social networks to deliver important goods and services. Drawing upon research undertaken within cities in the south and north of England, Williams and Windebank reveal that different kinds of informal work are undertaken within rich and poor areas. In higher-income neighbourhoods, paid informal work involved the use of firms and/or self-employed individuals, which were not paid formally for reasons of tax evasion. In lower-income neighbourhoods, meanwhile, informal paid work was largely conducted between friends, neighbours and other family members. Thus, Williams and Windebank argue that as much as one-third of all monetary exchanges in lower-income households were undertaken for reasons beyond the profit motive. Although money exchanged hands in these transactions, payments were made not so much to compensate for the labour and time expended, but rather to alleviate the recipients of the work from an obligation that this 'gift' would otherwise impose (Schrift, 1997). Thus, for Williams and Windebank, a good deal of work that goes on within certain communities is already 'alternative', in the sense that it is undertaken not so much for narrow, instrumental reasons, but is performed as part of much wider bonds of social reciprocity, and may be seen as a form of 'monetized mutual aid'.

The next chapter, by Colin Williams, Theresa Aldridge and Jane Tooke, looks at Local Exchange and Trading Systems (LETS), which in many ways attempt to develop an institutional means of regularizing and extending the mutual aid described in the previous chapter. LETS are examples of local currency systems through which participants can accumulate and expend money for a range of tasks and services. The chapter, which draws upon an extensive study of LETS within the United Kingdom, reveals that participation is skewed towards the socially excluded, and that people join for a range of reasons, both economic and non-economic. LETS are alternative economic spaces in at least two ways. First, they are alternatives to the formal economic sphere. Participation in LETS has, for some people, acted as a means of gaining skills and confidence that made it easier to them to move into the realm of formal employment. Secondly, LETS are also alternatives to the informal sphere. According to Williams, Aldridge and Tooke, LETS act as 'bridges', bringing together people who did not previously know each other. Thus, they can help bring the socially marginal and excluded into a broader network of social contact so that the possibility of participating within informal exchange of the kind described in the previous chapter becomes more likely. What LETS seek to do is formalize and 'scale' social networks within local communities so that the likelihood of reciprocal exchange is increased.

Chapter 8, by Jeffrey Jacob, looks at the North American 'back-to-the-land' movement of neo-yeomanism. This movement is driven by 'turnaround migration' and non-economic factors such as the desire to be closer to families and friends, and to achieve proximity to 'nature' and the assumed intimacy of small-town life. However, as Jacob points out, although the movement is also underpinned by an idealized notion of greater self-sufficiency through agricultural production, very few of the back-to-the-land migrants are able to carve out an alternative economic space that is capable of reproducing itself. Although the neo-yeomans surveyed by Jacob are able to produce about 30 per cent of their food needs from their properties, the rest of their food and other disbursements are supported by work necessarily undertaken within the formal economy. One of the reasons for this is that the 'back-to-the-land' social movement was driven by urban professionals who were rarely able to build up sufficient cash to be able to afford more than relatively small properties. As a result, almost half of the 'back-to-the-landers' surveyed by Jacob are effectively 'weekenders' who have to hold down other jobs due to the difficulties of carving out a self-sufficient existence. However, Jacob argues that the tide may be turning in favour of the neo-yeomans as environmental concerns about agricultural production mount. The small-scale nature of back-to-the-land operations means that they would be relatively immune to green taxes that sought to compensate for the environmental 'footprint' of agricultural operations. Thus, Jacob suggests that while it is difficult to sustain alternative agricultural spaces under current institutional and regulatory structures, a greater concern with environmental sustainability will make agricultural self-sufficiency a more viable possibility.

The chapters in this book, therefore, set out the possibilities, but also the difficulties, of carving out alternative economic spaces within an era of globalized capitalism. Indeed, the problems faced by many of the models and institutions outlined in this book might seem to confirm Harvey's pessimism for economic alternatives that seek to operate through a 'spatial fix', arguing that such attempts to escape the contradictions of capitalism will prove as temporary and fleeting as the those pursued by capitalist enterprises located much more firmly within the mainstream economy. What is needed, Harvey argues, is nothing less than a 'total' solution, through a set of social relations that produces an equitable and sustainable economic, political and environmental settlement (Harvey, 2000). Ultimately, of course, Harvey's pessimism may prove to be well founded. However, we would like to draw a more optimistic reading from these developments, and hope that the 'spaces of hope' outlined in this book are the antecedents of a more diverse, proliferative and inclusive economic future.

Note

1 However, Miller has subsequently argued that Callon's approach is supportive rather than critical of normative economics (Miller, 2002).

References

Bauman, Z. (1992) *Intimations of Postmodernity*. London: Routledge.
Bingham, N. and Thrift, N. (2000) 'Some new instructions for travellers: the geography of Bruno Latour and Michael Serres', in M. Crang and N. Thrift (eds), *Thinking Spaces*. London: Routledge. pp. 281–301.
Brenner, R. (2000) 'The boom and the bubble', *New Left Review*, II: 5–43.

Callon, M. (ed.) (1998) *The Laws of the Markets*. Oxford: Blackwell.
Carrier, J.G. (1998) 'Introduction', in J.G. Carrier and D. Miller (eds), *Virtualism: A New Political Economy*. Oxford: Berg. pp. 1–24.
Carrier, J.G. and Miller, D. (eds) (1998) *Virtualism: A New Political Economy*. Oxford: Berg.
Cassidy, J. (2002) *Dot.Con: The Greatest Story Ever Sold*. London: Allen Lane.
Cook, I. and Crang, P. (1996) 'The world on a plate – culinary culture, displacement and geographical knowledges', *Journal of Material Culture*, 1: 131–53.
Crook, C. (2001a) 'A different manifesto', *The Economist*, 27 September. Available at: http://www.economist.com
Crook, C. (2001b) 'Globalisation and its critics', *The Economist*, 27 September. Available at: http://www.economist.com
Derrida, J. (1994) *Spectres of Marx: The State of the Debt, the Work of Mourning, and the New International*. London: Routledge.
Dicken, P. and Hassler, M. (2000) 'Organizing the Indonesian clothing industry in the global economy: the role of business networks', *Environment and Planning A*, 32: 263–80.
Dicken, P., Kelly, P.F., Olds, K. and Yeung, H.W.-C. (2001) 'Chains and networks, territories and scales: towards a relational framework for analyzing the global economy', *Global Networks*, 1: 89–112.
Elliot, L. (2001) 'EC raises Tobin Tax hopes', *The Guardian*, 24 September. Available at: http://www.theguardian.com
Fannin, M., Fort, S., Marley, J., Miller, J. and Wright, S. (2000) 'The battle in Seattle: a response from local geographers in the midst of the WTO ministerial meetings', *Antipode*, 32: 215–21.
Ferguson, J. (1999) *Expectations of Modernity: Myths and Meanings of Urban Life on the Zambian Copperbelt*. Berkeley, CA: University of California Press.
Fine, B. (1999) 'The developmental state is dead – long live social capital?', *Development and Change*, 30: 1–19.
Fukuyama, F. (1989) 'The end of history?', *The National Interest*, 16: 3–18.
Gereffi, G. and Korzeniewicz, M. (eds) (1994) *Commodity Chains and Global Capitalism*. Westport, CT: Greenwood Press.
Gibson-Graham, J.K. (1996) *The End of Capitalism (As We Knew It): A Feminist Critique of Political Economy*. Oxford: Blackwell.
Gowan, P. (1995) 'Neo-liberal theory and practice for Eastern Europe', *New Left Review*, 3–60.
Gray, J. (1998) *False Dawn: The Delusions of Global Capitalism*. London: Granta.
Hamnett, C. (2000) 'The emperor's new theoretical clothes, or geography without origami', in G. Philo and D. Miller (eds), *Market Killing: What the Free Market Does and What Social Scientists Can Do About It*. Harlow: Pearson Education. pp. 158–69.
Hartwick, E.R. (1998) 'Geographies of consumption: a commodity chain approach', *Environment and Planning D: Society and Space*, 16: 423–37.
Hartwick, E.R. (2000) 'Towards a geographical politics of consumption', *Environment and Planning A*, 32: 1177–92.

Harvey, D. (1982) *The Limits to Capital*. Oxford: Blackwell.
Harvey, D. (2000) *Spaces of Hope*. Edinburgh: University of Edinburgh Press.
Hooper, J. (2001) 'Germany reacts to globalisation fears', *The Guardian*, 6 September, p. 13.
Hughes, A. (2000) 'Retailers, knowledges and changing commodity networks: the case of the cut flower trade', *Geoforum*, 31: 175–90.
Hutton, W. (1995) *The State We're In*. London: Jonathan Cape.
Jackson, P. (1999) 'Commodity cultures: the traffic in things', *Transactions of the Institute of British Geographers*, 24: 95–108.
Kelly, P.F., Olds, K. and Yeung, H.W.-C. (2001) 'Geographical perspectives on the Asian economic crisis', *Geoforum*, 32: vii–xii.
Klein, N. (2000) *No Logo: Taking Aim at the Brand Bullies*. London: Harper Collins.
Lee, R. (1989) 'Social relations and the geography of material life', in D. Gregory and R. Walford (eds), *Horizons in Human Geography*. Houndsmills: Macmillan. pp. 152–69.
Lee, R. (2002) 'Nice maps, shame about the theory: Thinking geographically about the economic', *Progress in Human Geography*, 26: 333–55.
Lewis, A. and Mackenzie, C. (2000a) 'Morals, money, ethical investing and economic psychology', *Human Relations*, 53: 179–91.
Lewis, A. and Mackenzie, C. (2000b) 'Support for investor activism among UK ethical investors', *Journal of Business Ethics*, 24: 215–22.
Martin, R. (1994) 'Stateless monies, global financial integration and national economic autonomy: the end of geography?', in S. Corbridge, N. Thrift and R. Martin (eds), *Money, Power and Space*. Oxford: Blackwell. pp. 253–78.
Martin, R. (1999) 'The new economic geography of money', in R. Martin (ed.), *Money and the Space Economy*. Chichester: Wiley. pp. 3–27.
Massey, D. (1995) 'Masculinity, dualisms and high-technology', *Transactions of the Institute of British Geographers*, 20: 487–99.
Mather, C. (1999) 'Agro-commodity chains, market power and territory: re-regulating South African citrus exports in the 1990s', *Geoforum*, 30: 61–70.
Miller, D. (1998) 'Conclusion: a theory of virtualism', in J.G. Carrier and D. Miller (eds), *Virtualism: A New Political Economy*. Oxford: Berg. pp. 187–215.
Miller, D. (2000) 'Virtualism: the culture of political economy', in I. Cook, D. Crouch, S. Naylor and J.R. Ryan (eds), *Cultural Turns/Geographical Turns: perspectives on cultural geography*. Harlow: Prentice Hall. pp. 196–213.
Miller, D. (2002) 'Turning Callon the right way up', *Economy and Society*, 31: 218–33.
Monbiot, G. (2000) *The Captive State: The Corporate Takeover of Britain*. Basingstoke: Macmillan.
O'Brien, R. (1991) *Global Financial Integration: The End of Geography*. London: Pinter.
Peck, J.A. and Tickell, A. (2002) 'Neoliberalizing space', *Antipode*, 34: 380–404.

Polanyi, K. (1957) *The Great Transformation*. Boston, MA: Beacon Press.
Ray, L. and Sayer, A. (1999) *Culture and Economy after the Cultural Turn*. London: Sage.
Reich, R. (2001) *The Future of Success: Work and Life in the New Economy*. London: Heinemann.
Robbins, P. (1999) 'Meat matters: cultural politics along the commodity chain', *Ecumene*, 6: 399–423.
Ross, A. (1997) *No Sweat*. London: Verso.
Sayer, A. (1995) *Radical Political Economy: A Critique*. Oxford: Blackwell.
Schor, J.B. (1992) *The Overworked American: The Unexpected Decline of Leisure*. New York: Basic Books.
Schrift, A.D. (1997) 'Introduction: why gift?', in A.D. Schrift (ed.), *The Logic of the Gift: Towards an Ethic of Generosity*. London: Routledge. pp. 1–22.
Smith, N. (2000) 'Global Seattle', *Environment and Planning D: Society and Space*, 18: 1–5.
Thrift, N. (1992) 'Light out of darkness: critical social theory in 1980s Britain', in P. Cloke (ed.), *Policy and Change in Thatcher's Britain*. London: Pergamon. pp. 1–32.
Thrift, N. (2000) 'Still life in nearly present time', *Body and Society*, 6: 34–57.
Thrift, N. (2002) 'The future of geography', *Geoforum*, 33: 291–8.
Wainwright, J., Prudham, S. and Glassman, J. (2000) 'The battles in Seattle: microgeographies of resistance and the challenge of building alternative futures', *Environment and Planning D: Society and Space*, 18: 5–13.
Winnett, A. and Lewis, A. (2000) '"You'd have to be green to invest in this": popular economic models, financial journalism, and ethical investment', *Journal of Economic Psychology*, 21: 319–39.

two

The Alterity of the Social Economy

Ash Amin, Angus Cameron and Ray Hudson

The social economy is increasingly seen as offering an alternative to the mainstream market economy and as a new model for restoring community and democratic participation. It is claimed to provide an ideal alternative for meeting social needs, particularly in areas where conventional sources of economic growth and social cohesion have been eroded as a consequence of private sector disinvestment and/or public sector cuts. In such areas, local social economy activities are proposed as a panacea, combining the provision of innovative welfare services, entrepreneurship, employment and training, the production of socially useful goods and services, and the strengthening of vulnerable communities. For many, the most appropriate institutions to achieve these goals are those understood to occupy the space between market and state, including not-for-profit organizations, social enterprises, community businesses and other third sector ventures.

The high expectations of the social economy have been pounced upon by policy-makers. Worldwide, many policy communities, including the European Commission (CEC, 1998a, 1998b, 2000), New Labour and the social democratic left elsewhere in Europe and North America (Gittell and Vidal, 1998; SEU, 1998) and international organizations such as the World Bank (World Bank, 1997) are promoting forms of local social enterprise as alternative means of developing social and economic capacities latent in even the most deprived communities.

But what does the 'alternative' offered by the social economy amount to and does it indeed offer the range of benefits expected of it in terms of employment, training, regeneration and community empowerment? Does the social economy as it is currently organized constitute a different and distinct means of organizing economic activity? Does it, as some academic commentators believe, represent not just an alternative way of managing aspects of a capitalist system, but the beginnings of an effective alternative to an alienated and profit-seeking capitalist praxis (Catterall et al., 1996; Ekins and Newby, 1998)?

The purpose of this chapter is to explore these questions in relation both to local social economy organizations in the United Kingdom and to the various policy programmes that have been introduced to develop and support them. After a brief introduction to the ways in which the concept of social economy as alterity has changed from the early nineteenth century to the present, the main body of the chapter considers the extent to which the local social economy in the UK can be seen to represent a real alternative to conventional economic activity in the public and private sectors. The discussion of the nature of the potential alternative offered by the social economy is organized around three main themes based on the most commonly stated policy expectations: the restoration of local community, self-reliance and needs-based service provision. By examining examples of social enterprise from throughout the UK we argue that, while examples of alternative processes and practices do exist, the experience of such projects casts doubt upon their capacity to deliver the range of alternatives expected of them. Rather, the nature of the alternative offered by the social economy is, in most cases, not only highly fragile and circumscribed by the maintenance of the mainstream economy, but is also dependent upon it. We argue in the conclusion that the constitution of the social economy as the main plank of local regeneration, runs the risk of offering an alternative that is just as iniquitous as that which it replaces, and that it is considerably weaker.

THE DEVELOPMENT OF SOCIAL ECONOMY AS A LOCAL 'ALTERNATIVE'

The concept of 'social economy', as an alternative way of organizing economic praxis such that its goals are more than the naked pursuit of profit, dates back at least to the early nineteenth century (Procacci, 1978). Until recently, although the precise ways in which conceptions of social economy were to be applied varied considerably, the various manifestations of the concept had at least one thing in common: they represented a form of alternative economic praxis that was to affect the (national) economy as a whole, reflecting in turn a conception of an associated social whole.

In this form, social economy represented an attempt by academic economists and politicians to effect the 'systematic grafting of morality onto economics' (Donzelot, quoted in Procacci, 1978: 61). The discourse of social economy was directly linked to the perceived problem of 'pauperism' during the early nineteenth century, a concept that denoted a profound moral laxity on the part of the urban poor rather than mere poverty; a laxity that threatened the cohesion and dynamism of the emerging industrial culture. The moral goals of social economy were therefore intended to 'correct' the deviant behaviour of paupers (through workhouses, pauper schools, prisons and asylums) as much as they were intended to alter industrial practice or redistribute wealth (Driver, 1993). Social economy did not represent an attempt to eliminate poverty, which was considered to be a natural and inevitable outcome of a normally functioning society (Bauman, 1998; Dean, 1991; Procacci, 1978). The 'alternative' that social economy represented took the form of a set of moralizing norms and philanthropic institutions imposed on the undeserving poor, rather than any reform of mainstream economic practice (Rose, 1998).

Later conceptions of social economy were similarly concerned to develop alternatives within the national economy as a whole, but were more radical in their agenda. Rather than seeking to develop limited alternative measures *within* the dominant free market version of capitalism, these conceptions, particularly as the concept was appropriated by the political left during the twentieth century, sought to develop alternative systemic forms of capitalism. According to Cassell (1923), social economy represented a systematic attempt to develop a more 'socialized' (if not socialist) form of national economy that recognized that it formed part of a social whole. As such, the concept is probably better thought of as a form of social *economics*, a modified form of capitalist economic practice, rather than as a prescription for *a* social economy, conceived as a specific, delineated domain within the capitalist economy operating on different rules to the mainstream. There were, of course, strong continuities between this conception of social economy and growing calls for greater regulation of the capitalist economy by, among others, Keynes (1933, 1936) and Polanyi (1957).

In part, therefore, this conception of the socialized economy ultimately became the basis for various systems of national social insurance, redistributive taxation, universal education and healthcare that formed the post-war welfare state. These various services, combined with industrial regulation, wages policies, the nationalization of strategic industries and other direct interventions in the functioning of the market, accorded to government the role of managing social goals directly in association with the management of the economy. Although not referred to explicitly as social economy, the welfare state undoubtedly represented the fulfilment of some of the earlier definitions of the concept.

However, over time the welfare state came under increasing pressure and criticism, particularly following the 'fiscal crisis of the state' first noted in the early 1970s, as its ability to deliver social progress and safety nets was significantly undermined (O'Connor, 1973). This encouraged various actors to attempt to restore the concept of the national social economy. For example, the 'Sheffield Group' of left-wing British academics and politicians came together at the end of the 1980s to attempt to reformulate the welfare and economic policies of the left against Thatcherite neo-liberalism (Alcock et al., 1989). Their conception of social economy was, again, a way of harnessing and transforming what would now be described as the 'mainstream economy' to achieve both immediate social goals and, in the longer term, a greater democratization of control that would 'entail on the one hand a re-fusion of social and economic policy geared to meeting human needs and, on the other, a more open, participatory and democratic state reflecting both the rights and duties of citizenship' (Alcock et al., 1989: 36–7). Advocating a form of socialized economic intervention, the Sheffield Group was explicitly concerned with the role of the state with regard to the national economy and, for all that it asserts the rights and duties of citizens, is equally emphatic about the central responsibility of the state to manage the economy in the interests of the social whole.

Through the 1990s, the concept of the social economy has continued to circulate, but in ways that differ markedly from many of these earlier conceptions, particularly in terms of the alternative they propose. There have been attempts on the conservative right to revive a nineteenth-century concept of social economy as a means of softening aspects of free market capitalism without raising the spectre of a renewal of the welfare state (Reisman, 1991). As with its antecedents, this notion of the social economy is not about developing an alternative to capitalism or eliminating poverty. Rather, the social economy as 'conservative capitalism' is intended to address a perceived contradiction for the right, between, on the one hand, an adherence to tradition (in the form of the nuclear family, the nation state, established religion, and so on) and, on the other, to the market as a force of dynamism and constant change (Reisman, 1999). Social economy in David Reisman's terms, then, is a radically deregulated market liberalism which, he argues, will automatically constrain the worst excesses of capitalist enterprise because the persistence of 'traditional' norms of behaviour will drive such practices from the marketplace.

However, the majority of recent accounts of social economy are concerned to develop a conception of *the* social economy, delivered through the third sector, which opens up a variety of alternative spaces and new institutional forms to meet specific social needs (see, for example, Anheier and Seibel, 1990; Borzaga and Santuari, 1998; Catterall et al., 1996; CEC, 1998a, 1998b, 2000; Gittel and Vidal, 1998; Grimes, 1997; Haughton, 1999; Henderson, 1999). As such, the nature of the alternative proposed by the social economy is significantly different from that in the earlier forms outlined above in at least two ways.

First, whereas all other conceptions of social economy have stressed the significance of a *socialized* economic praxis (for example, needs-based, participatory), *the* social economy is proposed as an economic circuit in its own right. As such, the social economy is seen to be distinct from the mainstream public and private sector economies, sometimes incorporating elements of both, but nonetheless standing as a distinctive 'third' sector (Anheier and Seibel, 1990). While this more recent concept of social economy continues to lay claim to alternative forms of economic praxis, these are not understood to be features that can or necessarily should extend to the other sectors of the economy. The social economy is still seen as an alternative to the mainstream, therefore, but importantly not one that replaces or reforms it.

Secondly, the social economy is routinely allocated to a specific spatial scale variously labelled the 'local', the 'community' or the 'neighbourhood'. The social economy is therefore understood to be an aggregation of localized third sector organizations. Although this is often presented as though it had a national form, as, for example, 'the UK social economy', this is distinctly different from earlier concepts of social economy understood as forms of economic praxis applicable at the national scale. Rather the 'national' and, more recently, following the establishment of regional assemblies in various parts of the UK, 'regional' social economies are agglomerations of localized practices that may have little in common other than being in some way distinct from conventional public or private sector organizations.

The assumption of local scale is a consequence in part of the widely accepted definition of social enterprises as being 'community-owned' and focused on providing for the needs of a specified local area (see, for example, Catterall et al., 1996; CEC, 1998a, 1998b, 2000; DETR, 1998; Ekins and Newby, 1998; Gittell and Vidal, 1998; Greffe, 1998; Haughton, 1999; Pearce, 1993; SEU,1998). The 'local' in this context, though rarely explicitly defined, usually refers to a small, definable territory and a homogeneous resident population – usually either a single housing estate, an established inner-city or suburban area or a rural village.

Local social economy organizations are understood to share a commitment to harness local economic activity and latent entrepreneurial capacity to create jobs and socially useful services by responding to the needs of the local community itself. So, for example, instead of local housing maintenance work being carried out by staff employed by a local authority or, as has been increasingly the case in the UK, by a private company on contract to the local authority, community-owned, non-profit enterprises are established within housing estates themselves to provide a more immediate and flexible service run by and for local people and to create jobs in the local economy (Saunders, 1997).

The social economy, therefore, is seen to stand as an alternative to the conventional national public and private sector economies in a variety of ways. First, it is claimed to offer an alternative means of strengthening local communities and boosting local economies. In having to worry about neither shareholders' expectations of profitability nor taxpayers' expectations of cost efficiency and lower local tax-rates, the local social economy is answerable only to the local community that owns and runs it. Secondly, the social economy demonstrates an alternative model of economic growth based on local social welfare rather than profitability. Thirdly, the social economy offers the possibility of a more holistic and democratic form of economic organization based on local participation and communal ownership. Finally, in having a strong link to the local community in which it operates, the social economy is expected to deliver social services in a qualitatively better way than has been the case in the past. In this way, the social economy acts as an alternative to a public sector 'nanny state' which is seen to have become overly bureaucratic, inflexible, expensive and to encourage dependency.

Overall, and where the current construction of *the* social economy overlaps most with the older forms of 'social economics', particularly those developed on the left, there is a common expectation that through growth and demonstration, the social economy can offer some form of alternative to market and state provision. This takes the form either of an alternative way of 'doing capitalism' – one that values social goals as well as capital accumulation – or a much more radical alternative to capitalism as a whole – the beginnings, however liminal, of a wholly different form of economic practice not driven by accumulation at all; some form of 'associative economy' (Archibugi, 1999) or 'egalitarian market' system (Bowles and Gintis, 1998).

The following sections analyse the various forms of alterity outlined above in relation to existing social enterprises in the UK[1].

STRENGTHENING LOCAL COMMUNITIES

The strongly local focus of much recent social economy debate and activity is seen as one of its most significant elements and one that distinguishes it from mainstream public and private sector economies constituted at a national level. The turn towards a localized form of social economy on the part of the policy community is underpinned by a desire to deliver a range of outcomes:

> ... both local and central state appear to want to provide conditional support for processes of community development in order to legitimise their wider political strategies, to improve local credibility, to reduce state expenditure and to create a culture of 'community responsibility' for regeneration. (Haughton, 1999: 7)

This is clearly expressed in, for example, the UK government's New Deal for Communities (NDC) programme, which is committed to developing 'very local' partnerships, including social enterprises. The NDC defines its target 'local' community in the following terms:

> The community selected should form a recognisable neighbourhood. It should not be so large that the partnership cannot focus its strategy effectively. Nor should it be so small that effective neighbourhood management strategies cannot be put in place. Communities will typically, therefore, be expected to cover between 1000 to 4000 households within a distinct area. (Haughton, 1999: 9)

While many academic commentators have criticized this communitarian turn as fracturing the principle of universal welfare provision (see, for example, Byrne, 1999; Frazer, 1999; Levitas, 1996, 1998; Rose, 1998), the localness of the social economy is seen by many as a means of reinvesting marginalized places with a community identity often thought to have been lost (CEC, 2000; SEL, 2000). Community-based interventions are, moreover, seen as valuable and necessary alternatives to failed bureaucratic, top-down approaches to regeneration (CEC, 2000; DETR, 1998; Haughton, 1999).

In practice the social economy comprises a huge variety of organizational forms including, for example, co-operatives, charities, companies limited by guarantee, credit unions, Local Exchange Trading Systems (LETS), voluntary organizations, tenants and residents groups, community associations, faith-based organizations, ethnic minority groups and all manner of informal social groupings. Although attempts have been made to confine the definition of the social economy to 'social enterprises', characterized by specific features concerning ownership, organization and practice (Anheier and Seibel, 1990; Borzaga and Santuari, 1998), in practice the criteria are much more fluid, encompassing much that would more generally be described as the 'voluntary sector' (see, for example, CEC, 2000).

There are certainly examples of social enterprises that do seem to provide evidence for the aspirations of the communitarian agenda. Among them are some of the best-known social enterprises in the UK – those that tend to appear with increasing regularity in the many best-practice studies that have been produced in recent years. These include, for example, the Arts Factory in South Wales, the Matson Neighbourhood Project in Gloucester and the Bromley-by-Bow Centre in the east end of London.[2] In all three cases, a mixture of commercial and social services have been brought into, or restored to, deprived communities, creating local jobs and improving living standards. In the case of the Matson and Bromley-by-Bow projects, both have successfully integrated the creation of new General Practitioner (GP) practices to improve local public health directly, with the provision of other services such as training and education, support for local shops, mental health provision and youth work. Both are situated in areas of primarily social housing which, until the projects were set up, suffered from a lack of basic local amenities, particularly accessible GP practices. The Arts Factory has developed a rather different set of services based around a wholesale and retail garden centre and plant nursery, a commercial pottery, public art projects and a multi-purpose community centre sited in a former church. The flexibility of the sites it has developed, along with the range of services offered (which include, for example, mental health support, training and youth work) means that the Arts Factory has become an important element in the local community.

In all three cases the contribution of the projects to the life of the local community has been marked. Both the Matson Neighbourhood Project and the Arts Factory have become champions for local people in relation to remote and bureaucratic local authorities. Both projects have, for example, with the strong support of local residents, successfully defeated unpopular and undemocratic planning proposals emanating from the local authority. As such, they have not only contributed to aspects of economic regeneration, but have successfully established new social, political and cultural networks as rallying points for local residents.

The ability of the social economy to develop this local capacity and to develop structures of local empowerment and democratization is not, however, universal. One of the longest established social economy projects in the UK, the Craigmillar Festival Society (CFS, founded in the early 1960s), despite being run largely by local people, has been found to have largely sidelined and ignored those local residents it was established to support. A recent report commissioned by the City of Edinburgh Council into the 37 community projects working in Craigmillar concluded that:

> ... the role of the CFS as the 'voice of the people' should be challenged. The perception was that the attendees at the general meetings by which the CFS executive asserted their democratic mandate were generally project staff who had a vested interest in maintaining the status quo. (DTZ Pieda Consulting, 1999: 65)

The *status quo* consisted, to a large extent, of the 'community'-owned initiative failing to involve the community of Craigmillar and Greater Niddrie in any decisions over the needs and priorities of the local area.[3] As a consequence of this report, the umbrella organization for the many community projects operating in the area has been wound up on suspicion of the misappropriation of funds and the remaining projects, including CFS, rationalized and subject to funding cuts.

The scale of the problems facing the Craigmillar area was such that the development of any form of democratic regeneration, based largely or wholly upon the mobilization of local capacity, was never going to be easy. However, in the case of CFS, which has long been considered a success story of the social economy, the result of its failure to address the needs of the local community has been a considerable reduction in the services available in the area. Above all, despite its long track record of project development, CFS notably failed to develop any alternative social, political or economic structure for local people outside a small coterie involved with the running of the project itself. The result was a pervasive mistrust of the project among local residents who felt, rightly or wrongly, that project leaders were more concerned to develop their own power bases in the area than to address the needs of the people.

Although CFS is perhaps an extreme example, there are many others where, even though the project in question is genuinely inclusive and offers an improved form of local service, there is considerable hostility from sections of the community that it serves. One area of significant and growing conflict, for example, has been in relation to the transfer of large blocks of social housing from local authority control to the social economy (Hetherington, 2000). For example, one of Bromley-by-Bow's partner organizations in Tower Hamlets the Poplar HARCA,[4] was established in 1998 to manage 11 blocks of housing in the east of Tower Hamlets containing over 7,000 homes. Although the HARCA plans to develop a range of integrated and holistic social and environmental services, and although rents are now lower than those of the local authority, the project has been beset by conflict. A local campaign opposing the 'privatization' of public housing stock, even if to the third sector, has prevented a planned third phase of housing coming under the HARCA's control.[5] The campaigners, rightly or wrongly, succeeded in thwarting the transfer by playing upon local ethnic, cultural and political differences to convince many residents that the transfer to the HARCA would not be in their interests. They have argued, for example, that rents would rise very quickly as the project's private sector lenders – a consortium of commercial banks – called in their debts, and that levels of maintenance and repair would fall. Paradoxically, although the HARCA's constitution prevents either of these things from happening, a sufficiently large minority of the local community voted against the third phase of the transfer to prevent it. This example illustrates how problematic the idea of local community remains in areas already divided on racial, class and other grounds.

A strongly local focus for the activities of social enterprises can also be a drawback because of the limited capacity for expansion in deprived communities. Where clusters of local social economy projects develop in particular regions, such as in the former coalfields of South Wales, for example, the competitive nature of social economy funding, the limited range of activities undertaken by social enterprises and the limited capacity of the depressed local market, means that in practice the expansion of the local social economy is impossible. For that reason some exponents of the social economy have argued against the inward-looking focus of many community-based organizations, which tends to produce what is in effect a 'ghetto' economy, rather than an outward-looking local economy able to tap into wider, non-local networks and markets (see, for example, West, 1999). Even where such networks can be established, however, the evidence that social enterprises have the capacity to compete on an equal footing with private sector enterprises is ambiguous (see, for example, the case of Tayside Community Business in Hayton, 2000).

In light of this problem of limited local capacity (local here in the sense of neighbourhood as defined by the DETR in the quote above), many social economy organizations are sited at much larger spatial scales. On Teesside, for example, although there are a number of small and rather precarious community-based projects, the most notable success stories in the local social economy have been those operating on a city-wide scale. One of the most prominent of these is the Community Campus '87 project for people with special housing needs, which provides a mixture of sheltered housing, holistic social support and work training, using the refurbishment of its properties to provide work experience for its tenants. The work of the tenants increases the housing stock owned by the project which, in turn, provides more training places to fund and refurbish more properties. Community Campus is able to grow in this way because of the very depressed local housing market, allowing it to buy empty and often derelict buildings very cheaply. It could not, however, operate at the scale of a single housing estate or 'neighbourhood' (DETR, 1998), partly because the problem it seeks to address is not confined to a specific area, and partly because it needs the economies of scale of the wider area to deliver the alternative service that it provides.

Similar dynamics drive a few social enterprises that, despite often being described as local, are effectively detached from any identifiable area or community operating on multiple spatial scales up to and including the national level. The Wise Group in Glasgow, for example, which is probably the social enterprise in the UK most cited as an example of best practice (see, for example, CEC, 1998a, 2000; DETR, 1998), was only for a short time in the early 1980s oriented towards local neighbourhoods. When it began, the project's services (providing insulation, security devices and landscaping in areas of social housing combined with training) were delivered by community-based 'squads' dotted throughout Glasgow and responsible for particular areas and/or communities. As the project has grown, however, the Wise intermediate labour market in Glasgow has been organized on a city-wide basis, with all operations being based in the project's Charlotte Street offices. In addition, Wise has developed a number of subsidiaries and associated projects in cities throughout the UK which, although more or less local depending on the nature of the local labour market, follow a model developed elsewhere. Since its inception, Wise has followed its chairman's oft-stated belief that if something is to be effective it has to be done 'at scale'. More recently, following changes in the nature of employment services and Scottish devolution, Wise has been taking on an increasingly regional and even national role. Not only has Wise been bidding to run one of the government's pilot regional Single Work Access Gateways (Clyde Coast), but it is looking to expand as a provider of training and work experience throughout the UK.[6]

As these various examples illustrate, the issue of locality with regard to the social economy is, despite its ubiquity in the policy literature, ambiguous. First, the degree to which social enterprises relate to the local areas they are intended to serve varies enormously, ranging between identifiable and self-defining communities up to city- and region-wide catchments. In some cases the 'community' served may be a specific interest group (youth, the homeless, the learning disabled, for example) rather than a single, territorially defined population. In other words, although all social economy organizations are expected to be local, some are considerably more local than others.

Secondly, the extent to which the 'localness' of the social economy is a product of local factors (for example, local needs identified and acted upon by local people), rather than the exigencies of regional or national policy programmes, needs to be considered. While the social economy is expected to tap into and/or constitute some form of 'authentic' local capacity as the basis for the alternative it offers, the prescriptive nature of much public policy might suggest another reading. If a community focus arises more out of the demands and expectations of national policy than from existing structures and capacities in local communities, to what extent can this be said to be local?

Thirdly, this is particularly significant given the relationship between alterity and locality. Specifically, it is important to make a careful distinction between, on the one hand, 'local' as a site and scale at which alterity is developed and expressed and, on the other hand, the presentation of the 'local' as the alternative itself. If the local is the site at which alterity is expressed, as is certainly the case with the first three project examples given above, then the nature of the local is fluid and inclusive. The significance of locality for these projects is not that it delineates and defines a bounded space within which social economy activities are contained (the local can refer as much to a city as a neighbourhood), but a point of commonality that serves to unite people as local. The local community in such circumstances effectively becomes an alternative centre (alternative to, for example, a local authority) around which other activities develop.

In the latter sense – the notion that locality *is* the alternative – the nature of the local is more problematic. Importantly, as illustrated all too clearly by the quote from the New Deal for Communities document above, such a prescriptive concept of the local demands that it be defined as a quantifiable and bounded space. If locality is in itself the alternative (in other words, that the local scale is promoted as the appropriate alternative scale for welfare provision), then it is essential that the 'local' be definable in order that it can be replicable. It is for this reason that, as welfare policy has come to embrace the local as the most appropriate and, increasingly, only site of intervention, so considerable efforts have been made to quantify local spaces. Notable in this process, for example, in the UK context has been the Centre for the Analysis of Social Exclusion (CASE) at the London School of Economics, established in 1997. CASE has worked closely with the Social Exclusion Unit (SEU) in defining a highly localized welfare agenda often prescribing social economy solutions (see, for example, Glennerster et al., 1999; Power and Bergin, 1999; Smith, 1999). This in turn also contributes to a similar dialogue between academic research and policy-making in relation to European Commission policy (CEC, 1998a, 1998b, 2000). In both sets of literature, the 'local' is defined vaguely and indeterminately, loosely linked by association with the concepts of community and neighbourhood. In practice, a new meaning for the 'local' is being generated through such policy discourses which equates localness with poverty and exclusion, as well as institutional responsibility for dealing with the latter. As Ruth Levitas puts it with deliberate irony, in the new welfare policy, 'community rules' (Levitas, 1998: 89).

SELF-RELIANCE

Much support for the social economy is predicated on the belief that social enterprises can be self-funding in the longer term and that they can develop capital assets for the local community. The emphasis on employment creation and entrepreneurship in the social economy is not just understood to be the best way of tackling poverty, but is often presented by the policy literature as the only way of pursuing the objective of making local communities economically self-sufficient (CEC, 1998a, 2000). The New Deal for Communities, for example, is quite explicit that it will only consider projects that 'have a clear forward strategy which explains how the partnership will keep the momentum going beyond [the] ten year period [of funding]' (DETR, 1998: 18).

In practice only a very small number manage to free themselves of grant funding completely. A 1997–98 study for the European Commission's LOCIN project, revealed that 72 per cent of 173 UK projects established for at least three years were still wholly reliant on public funds while a further 16 per cent relied on the public purse for at least 70 per cent of their income (Amin et al., 1998). Only 3 per cent could report that they were wholly free of public sector grants for revenue funding, but even among these projects, some made sporadic use of grants for new project development. Of course, sustainability does not necessarily imply that projects must become wholly self-financing. Rather it implies a balancing of sources of income such that the proportion of grant income tends to fall over time so that such limited funds as are available are used to best effect. However, in the context of falling welfare budgets, the question has to be asked if the financial autonomy expected of the social economy is simply a substitute for state welfare expenditure rather than a genuine addition to the resources available for tackling poverty and exclusion.

As indicated above, the numbers of projects that can be said to be in any way financially sustainable are relatively few in number. That said, there are important examples where innovative development strategies, creative use of public sector sources and often fortuitous and contingent local conditions have enabled social economy projects to develop significant income streams and assets.

Although founded with a range of public sector grants, the Spitalfields Small Business Association (SSBA) has been almost wholly self-financing since the mid-1980s. Since its inception in 1979, SSBA has acquired, through various means, a large portfolio of property in the Spitalfields area of east London. Some of this property is leased to a housing co-operative intended originally to improve housing conditions for the local Bangladeshi and other ethnic minority communities, while the remainder is leased to local small businesses by SSBA as managed workspace along with the supply of a range of business development, training and consultancy services. SSBA currently manages over 65,000 square feet of workspace, housing around 60 small and medium enterprises. Of this 45 per cent is owned freehold, amounting to a significant community-owned asset in an area in which property values have soared in recent years. Of its annual income of £750,000, 46 per cent is derived from workspace rents, 28 per cent from an ERDF business support programme, and the remainder from a mixture of fees, charitable donations and various public sector contracts. By controlling rent rises carefully, SSBA has also insulated vulnerable local businesses from the very high rises in commercial rents in the area during the 1980s and 1990s.

In other cases, rather than developing assets and income over a period of time, projects have been created specifically to manage and further develop existing assets. Coin Street Community Builders (CSCB) on the south bank of the Thames was created to act as a management and development company for a 14-acre site in central London after the local community won a long campaign to prevent the area being sold to commercial developers for office buildings. The Greater London Council sold the site to CSCB for a fraction of its true value and protected its status as a community-owned asset through a series of binding covenants. The result has been the creation of a varied development scheme embracing tenant-owned and managed social housing schemes, workspaces for private sector designers, retail outlets and catering, exhibition spaces, public parks, training and employment services and an annual community arts festival.

For the bulk of the social economy, however, the major source of income remains the public sector, either in the form of grants or service contracts, both of which serve to constrain rather than enhance the possibilities for growth. Even where surplus could be created by social enterprises there are often positive disincentives to do so. Some funders, for example, will claw back surplus funds at the end of the funding period (SEL, 2000). Not only are funders notoriously 'risk averse', but many are formally prevented from funding any organization that explicitly intends to generate a surplus. National Lottery funds, for example, cannot be used to support social enterprises which are established to create a surplus, even though it is not distributed as profit (SEL, 2000).

The availability of most funds on an annual basis only, changing criteria for eligibility, onerous administrative and supervision requirements and a host of particular problems (such as retrospective and often late payment by the European Social Fund) makes medium- to long-term financial planning impossible. Take the case of Pecan Ltd., a highly successful employment training project in south London. Project leaders must produce sets of accounts for four different financial years, each presented differently, while the project can be audited without warning by any of 14 different funding bodies. Despite the widely acknowledged success of the initiative in tackling local unemployment, and the extreme administrative load, almost all of Pecan's funding is annual. The sort of long-term financial management that might help social enterprises to become self-financing is therefore rendered practically impossible.

For the majority of social enterprises in the UK, the problem is not how to manage assets, but how to acquire them at all. Those projects that have managed to develop assets over time have been able to do so because of a degree of continuity in their financing, or because the activities in which they are engaged are intrinsically accumulative – especially those based on property. In some cases, this involves the development of alternative ways of combining income from a variety of sources (usually the public sector plus revenue from rents or contracts), which in a minority of cases, such as those cited above, can lead to the creation of viable social enterprises and substantial assets. However, while such examples may represent alternative means of creating income and assets, this is often in practice a different way of acquiring resources rather than a qualitative difference in the process of accumulation. The need to find such different means is often less a product of any ideological commitment to the social economy than a product of necessity because of the restrictions associated with regeneration funding. The short-termism of many funds, rapidly changing funding criteria, constraints on access to private loan capital, the limited nature of local market capacity, and so on, mean that social enterprises have no choice but to be 'alternative' in the way they stitch together different parcels of funds. Alterity in this sense may, as we have seen, produce some startling results, but it is a product not of the independence of the social economy from the state, but rather a product of its ongoing dependence upon it.

To return to the more general issue of whether the social economy is expected to offer alternative ways of maximizing the benefit of public sector funds or an alternative *to* public sector funds, the latter is increasingly the case *de facto* as regeneration programmes are scaled back and/or spread ever more thinly. If one contrasts, for example, the sums spent on regeneration in the East End of Glasgow over the past 30 years the figures are stark. During the Glasgow East Area Regeneration (GEAR) programme, which was launched in the mid-1970s, as much as £60 million a year was spent in a small area of the East End every year for eight years. Huge cutbacks in local authority spending as a consequence of a 'fiscal crisis' caused by local government reorganization in 1996 (Carmichael and Midwinter, 1999), as well as radical reform of Urban Programme funding mean that the current Social Inclusion Partnership (covering a larger portion of the same part of the city) had just £2.8 million to spend on a wide range of projects during 1998–99 (Turok and Hopkins, 1997). The availability of even these meagre funds was contingent upon meeting stringent output targets set by the funding authorities, and subject to competition from the whole of Scotland (Turok and Hopkins, 1997).

In such circumstances, the local social economy, although touted as a means of preserving and extending public monies, in practice is expected in the longer term to be an alternative to the public purse financing local welfare provision. The capacity for the social economy to provide such an alternative at any time is, as indicated above, likely to remain very limited, even if some of the constraints imposed by the prevailing funding regime could be eliminated.

MEETING NEEDS

In the area of service and welfare provision, in practice the vast majority of social economy organizations is engaged in activities that would previously have been the preserve of statutory authorities. This is partly a consequence of constraints imposed by funders, particularly during the 1980s when resources could not be made available to any organization that would be in competition with local private sector businesses. It is also partly a consequence of the main driving force behind social enterprise development over the past two decades, which has been the privatization and/or contracting out of local authority services.

The nature of alterity with regard to service provision is not merely that the social economy takes over areas of public sector provision as is. Rather, there is a hope and belief that the alternative services offered by the third sector will be qualitatively better than its public sector predecessors. Of the larger-scale projects, some have, in effect, become alternative forms of local state institution in their own right, taking over almost complete control of significant areas of service provision. In the case of Poplar HARCA cited above, the organization has developed a range of services, including training, childcare, maintenance and cleaning services, health promotion and general housing management, to such an extent that it has become a form of quasi-local authority for those tenants in the housing it controls. Part of the resistance to the development of the HARCA has been specifically in relation to opposition to the scale of devolution of statutory powers that this process seems to entail, and, consequently, fears over accountability. Although there is clear evidence that the services offered by the HARCA to its tenants are significantly better than those offered by the Tower Hamlets council, at least in the short term, and despite a long history of very poor services and under-investment in the east of the borough, the alternative has yet to be 'sold' effectively to a large minority of local people.

Elsewhere, social enterprises are able to create entirely new services, rather than alternatives to existing provision, and deliver them to particular sections of the community. The Reclaim project in Sheffield and the Gabalfa Community Workshop in Cardiff have both, albeit in different ways, pioneered the provision of work experience and training for people with severe learning difficulties. Reclaim is a plastics reclamation business that runs one of the few plants in the UK to sort and prepare the full range of plastic materials and textiles for recycling. It was established as a flagship environmental project in 1989 when Sheffield was nominated as Friends of the Earth's 'Recycling City of the Year'. Reclaim was able to offer unique work experience for a particularly marginalized group in the city, for whom, until that time, there had been no form of work training available. Despite enormous difficulties in finding and sustaining funding, Reclaim has helped a number of its clients into work, either through employment at its own factory or in other organizations in Sheffield. Gabalfa was established to fill a similar gap in provision for people with learning disabilities in Cardiff. The project has been designed to help such people gain sufficient skills and confidence to enable them to enter the mainstream labour market from which they were previously systematically excluded. As in Sheffield, until Gabalfa started, services for people with learning disabilities were confined to day-care centres or low-level forms of occupational therapy that were designed to cope with existing problems rather than to develop new abilities. Both projects have developed into highly successful social enterprises which, while in neither case being wholly independent of the state, have created very valuable, alternative local services.

However, even where local social economy organizations are able to develop alternative welfare services that do much more than merely replace public sector provision, their existence and continuity are often extremely precarious. In the case of Reclaim, the project was almost destroyed when a small annual grant from the British Plastics Federation was summarily withdrawn after three years. Although this represented a fairly small proportion of the project's income at the time, no other funders were willing or able to fill the shortfall completely, resulting in a deficit which took Reclaim over two years to resolve. Had Reclaim been unable to replace the lost funds, as was very nearly the case, the service that it provides for a particularly needy group in Sheffield would simply have been lost. Such stories are not uncommon.

Although in a few places, local authorities are prepared to underwrite certain funds to provide projects with a degree of continuity and stability, this is by no means the norm and does not in itself guarantee the continuity of non-statutory services. In the majority of cases, therefore, the alternative services that can be provided through the social economy are unable to benefit from the relative security of public sector provision underwritten by the state. The flexibility and innovation of social economy based-services, in other words, are all too often being bought at the cost of security and continuity. When the collapse or withdrawal of such services affects the most vulnerable groups in society, the value of such an alternative has to be in question.

WHAT KIND OF ALTERITY?

In light of the discussion above, how can we interpret what is alternative about the social economy in the UK? What is it an alternative to and for whom?

The conventional wisdom concerning the nature of the alternative represented by the social economy is that it is, or should be, an alternative to both private and public sector praxis. However, as we have noted above, current experience shows that the social economy remains highly dependent on the public sector for its continuity and development and, given some of the constraints that it faces in terms of developing income and assets, seems likely to remain so for the foreseeable future. This implies that rather than being an alternative to public sector provision, the social economy represents an alternative way of organizing such provision. Services are still provided to needy communities which might formerly have been provided by the state, and these may be new or better services than was previously the case, but the more fundamental point is that the nature of the relationship between the state and the (poor and marginalized) citizen has significantly altered. As we have argued elsewhere (Amin et al., 1999, 2002), the local social economy involves a significant shift of risk away from the state (national and local) and on to those sections of the population least able to absorb it. As such, the localization of welfare and regeneration activities represents a significant fragmentation of the universal principle of the national welfare state and its devolution into organizations that are in many cases under-funded, highly dependent and extremely precarious.

The alternative that the social economy is expected to represent with regard to the private sector is similarly problematic. Only rarely does the social economy directly replace or compete with the private sector as the provider of goods and services in the way that it does with the public sector. This is in part a consequence of the nature of the scale on which the social economy is expected to operate – the local – and the close association of the third sector with poor communities. In most cases this means that the social economy provides alternatives to the private sector not because that provision is necessarily better, although this may be the case, but because the social economy as currently constituted is expected to fill spaces *abandoned* by the private sector. This will, for example, often take the form of a local social enterprise taking over retail outlets to ensure a local food supply or to market locally produced or recycled manufactured goods. While such efforts are often of value, given the restricted nature of local demand, the potential for growth is severely constrained, not least because social enterprises have inadequate means for developing new and expanding markets. More importantly, the local social economy appears less as a positive alternative to private sector praxis, than as the only alternative available to mitigate the effects of private sector disinvestment. Effectively it means the semi-socialization (given the pervasive dependence of the social economy on the public sector) of activities previously provided by the private sector, with none of the costs or risks associated with this process borne by the private sector other than in the form of entirely voluntary philanthropy.

Many social economy organizations, for example, offer training courses to local unemployed people or to women returners, which are specifically designed to equip them for absorption into the mainstream, usually private sector, economy. While the social enterprise providing the training may represent an alternative form of praxis, at the same time it relieves the private sector of some of the significant costs and risks associated with training and recruitment. This may help to attract inward investment and protect local jobs by lowering costs for and protecting the profitability of local firms, but it can hardly be read as an alternative to the mainstream when, again, the social economy is subordinate to and contingent on the mainstream sectors.

These ambiguities clearly raise the issue of who is expected to benefit from the alternative represented by the local social economy. The assumed beneficiaries of the localization of welfare provision and economic regeneration are local people themselves, who are expected to gain from better services, new job opportunities and strengthened communities. However, not only is this by no means guaranteed, as in the case of CFS above, but there are other beneficiaries too. As suggested above, the state, local and national, benefits from the devolution of the risk and responsibility to the local level and from the generally positive messages concerning inclusion and accountability that accompany it. Similarly, the private sector benefits indirectly from the training and other supply-side and infrastructural services that the social economy provides. There is, however, another and growing group of beneficiaries who perhaps gain the most from the local social economy – the professional 'social entrepreneurs' who in many cases run it.

Many social enterprises are established and organized by people whose commitment to their cause is such that they are prepared to work long hours for little material reward in pursuit of their social and community goals. Salaries for social economy staff are usually much lower than comparable public or private sector staff, and in a few cases where the enterprise is organized as a workers' collective, all staff, from cleaners to managers, receive the same wage. Many people work in the social economy primarily because of a social, religious or political commitment rather than for the salary and, as such, provide a subsidy in kind to their organizations over and above their contributions in terms of skill and labour. Such individuals, through their strong personal commitment and personal praxis, offer a clear alternative to conventional attitudes to work based on income maximization and personal aggrandizement.

However, as the concept of the social enterprise has become increasingly the focus of public policy, so a class of 'professional' social entrepreneurs, who, in effect, 'deliver' the local social economy on behalf of local people, has developed. During the 1970s and 1980s, social economy organizations tended to be a product of committed people responding to the needs of the communities in which they lived, whether or not they originated in that community themselves. As the social economy has increasingly become the object of public policy, however, and particularly in the UK since 1997, so social entrepreneurship has become increasingly detached from place. In addition to the creation of a number of university and college courses in community regeneration and social entrepreneurship, there has been a proliferation of intermediary, advocacy, consultancy and co-ordination organizations designed to promote the social economy as a coherent and professionalized activity.[7] In effect, this is producing a class of social economy professionals who move from place to place and project to project 'fixing' local problems and importing an ideology of community which may have little or no connection with the place in which they operate.[8] Such issues are by no means lost on the 'local' people for whom social economy solutions are prescribed. Women running a project on the Isle of Dogs, for example, all of whom were local residents, disparagingly described such social economy professionals as 'the poverty pimps'. Many of the jobs created in the social economy, particularly the more senior ones, are held by such professionals. This may in many cases be necessary because of local skills deficits, but as the social economy takes on the appearance of an alternative 'regeneration industry', there is a danger that it will become increasingly bureaucratized and standardized – a 'one-size-fits-all' solution applied to 'communities' and 'neighbourhoods' by 'experts'.

CONCLUSION: THE ELISION OF THE LOCAL AND THE SOCIAL

In his account of the 'Third Way', Anthony Giddens is very clear about the significance of the 'community' in the new order:

> The theme of community is fundamental to the new politics, but not just an abstract slogan. The advance of globalization makes a community focus both necessary and possible, because of the downward pressure it exerts. 'Community' doesn't imply trying to recapture lost forms of local solidarity; it refers to practical means of furthering the social and material refurbishment of neighbourhoods, towns and larger local areas. (Giddens, 1998: 79)

This notion that community can be (re-)made through the mobilization of local capacity lies at the heart of the current growth of interest in the local social economy and is fundamental to the welfare reform programmes of the UK and other western governments and the European Commission. As we have demonstrated above, however, the evidence from the social economy itself is extremely ambiguous and uneven, questioning the extent to which it can really be seen as a 'practical means' of delivering alternative forms of social and economic regeneration.

At one level, current policy measures retain echoes of some of the earlier attempts at formulating a social economy praxis by, in Donzelot's terms, 'grafting morality on to economics' – producing an alternative economics whereby the needs of the social collectivity are incorporated into economic practices at all levels. The fundamental difference, of course, is that the most recent manifestation of social economy does not apply to all levels of economic practice at a national scale, but effectively only to those localities, communities and neighbourhoods where conventional economic practice is deemed to be no longer possible, practical or desirable. The 'social' in the social economy, thereby becomes equated not only with the local, but with particular categories of the local characterized by poverty and 'exclusion'.

This suggests that the pursuit of the social economy as a vehicle for regeneration may not result in genuine economic innovation and the 'socialized' economy, but, to paraphrase Finn Bowring, to the development of a 'second class economy' (1998: 106). The concentration of such initiatives in areas already both physically and functionally separate from the core of the mainstream economy may well have the tendency to *reinforce* existing geographies of uneven development. As Bowring notes with regard to LETS projects (themselves being widely promoted as local regeneration models (for example, CEC, 2000)):

> As LETS are held out to the unemployed, in true Victorian spirit, as a genuine opportunity to better themselves, the abandonment by mainstream society of the jobless poor, and of the welfare services they depend on, will be legitimized, and their exclusion from the structural guarantors of social identity and citizenship consolidated. (Bowring, 1998: 107)

The emphasis placed upon entrepreneurial solutions to local needs and, indeed, the explicit situation of these various projects and organizations in a social *economy*, enhances this separation between the inclusive nature of the Keynesian welfare state and the fragmented and selective nature of what Bob Jessop has described as the 'Post-national Schumpeterian Workfare State' (Jessop, 1993, 1995).

Notwithstanding the many examples of inclusive and democratic community organizations throughout the UK, they cannot be considered to constitute an alternative 'social economy'. As we have argued above, the nature of the alternative that they are expected to develop is contradictory. Increasingly, social economy organizations are expected to provide an alternative form of localized welfare and regeneration that simply replaces older forms of welfare provision with more precarious, weaker and uneven forms. This is not, it should be stressed, intended as a criticism of the projects themselves, except in those cases such as in Craigmillar where local projects have clearly misrepresented or ignored the local community. Rather, it is a criticism of a growing policy consensus that sees in the devolution of responsibility and risk away from the state a means of delivering services in a manner that protects both state and market from 'undue' demands on their resources. Moreover, as Hayton notes in relation to the failed 'community business' scheme that ran in Scotland during the 1980s and early 1990s, localized social economy experiments are often funded on the basis of 'activists' rhetoric', rather than on any very solid evidence that such projects actually work (Hayton, 2000: 204).

In the *Manifesto of the Communist Party*, Marx and Engels wrote scathingly of a particular form of 'conservative' or 'bourgeois' socialism which sought to mitigate the effects of capitalism without actually replacing it:

> [Bourgeois] Socialism, however, by no means understands the abolition of the bourgeois relations of production ... but administrative reforms, based on the continued existence of these relations; reforms, therefore, that in no respect affect the relations between capital and labour, but, at best, lessen the cost, and simplify the administrative work, of bourgeois government. (Marx and Engels, [1872] 1967: 114)

In many ways, the manner in which local social economy policy has developed in recent years, and particularly the way in which it has been embraced so wholeheartedly by New Labour and other advocates of the Third Way, constitutes just such a process of 'administrative reform'. Unfortunately for the few social enterprises that predate the current fashion for the social economy, and which have made extraordinary contributions to the lives of the people and places that they serve, their work has been co-opted into a policy discourse that is more concerned to provide more efficient (that is, 'cost-effective') welfare, than the sort of radical alternative that some of them indeed represent.

Notes

1 This chapter draws on the findings of two major studies of social economy activities conducted by the authors at Durham University. The first gathered data on 173 social enterprises of various forms for the European Commission's best practice study – *Local Initiatives to Combat Social Exclusion in Europe* (LOCIN) – during 1997 and 1998. The second project, *Constructing the Social Economy through Local Community Initiatives?* (ESRC No. R000237967), examined a smaller number of projects in four urban areas in the UK (Glasgow, Bristol, Middlesbrough and Tower Hamlets) during 1999 and 2000, to assess the significance of local factors in determining the relative effectiveness of social economy development in different places. A more developed account of this research can be found in Amin et al. (2002).

2 All three projects have been cited as examples of best practice in various policy documents (see, for example, SEL, 2000 and DETR, 1998).

3 Articles covering the so-called 'Craigmillar affair' appear in the following editions of the *Edinburgh Evening News*: 1998 – 6, 7, 10, 11, 20 and 30 November, 1, 26 and 31 December. 1999 – 4, 8, 14, 15, 21, 23 and 30 January, 3 February, 6, 9, 11, 12 and 16 March and 11 June.
4 Housing and Regeneration Community Association.
5 Similar campaigns have been mounted in other UK cities, most notably Glasgow and Birmingham, where similar attempts to transfer housing stock to third or private sector housing associations are under way (Hetherington, 2000).
6 Interview with the Wise Group, Glasgow, 8 June 1999.
7 One of the most recent and most influential of these, for example, is Social Enterprise London (SEL), established in 1998 to lobby the new London Assembly on behalf of the social economy for the whole of London. SEL has close links to the Treasury and has been closely involved with the Social Exclusion Unit's 'Policy Action Teams' (see http://www.sel.org.uk) (SEL, 2000).
8 Asked about the fact that so few 'local' projects in Wales were run by local people, for example, one project leader commented that 'the Welsh are not very good at community' (interview, 23 February 1998).

References

Alcock, P., Gamble, A., Gough, I., Lee, P. and Walker, A. (eds) (1989) *The Social Economy and the Democratic State: A New Policy Agenda for the 1990s*. London: Lawrence and Wishart.

Amin, A., Cameron, A. and Hudson, R. (1998) 'Local initiatives to combat social exclusion in Europe: final report'. Unpublished research report to the European Commission *LOCIN* Project.

Amin, A., Cameron, A. and Hudson, R. (1999) 'Welfare as work? The potential of the UK social economy', *Environment and Planning A*, 31: 2033–51.

Amin, A., Cameron, A. and Hudson, R. (2002) *Placing the Social Economy*. London. Routledge.

Anheier, H.K. and Seibel, W. (eds) (1990) *The Third Sector: Comparative Studies of Nonprofit Organizations*. Berlin/New York: Walter de Gruyter.

Archibugi, F. (1999) *The Associative Economy: Insights Beyond the Welfare System and into Post-Capitalism*. Basingstoke: Macmillan.

Bauman, Z. (1998) *Work, Consumerism and the New Poor*. Buckingham: Open University Press.

Borzaga, C. and Santuari, A. (eds) (1998) *Social Enterprises and New Employment in Europe*. Trento: Regione Autonoma Trentino–Alto Adige/European Commission DG5.

Bowles, S. and Gintis, H. (1998) *Recasting Egalitarianism: New Rules for Communities, States and Markets*. London: Verso.

Bowring, F. (1998) 'LETS: An Eco-Socialist Alternative?', *New Left Review*, (Nov.–Dec.): 91–111.

Byrne, D. (1999) *Social Exclusion*. Buckingham: Open University Press.

Carmichael, P. and Midwinter, A. (1999) 'Glasgow: anatomy of a fiscal crisis', *Local Government Studies*, 25 (1) (Spring): 84–98.

Cassell, G. (1923) *The Theory of Social Economy*. London: T. Fisher Unwin.

Catterall, B., Lipietz, A., Hutton, W. and Girardet, H. (1996) 'The Third Sector, Urban Regeneration and the Stakeholder', *City*, 5–6: 86–97.

CEC (1998a) *The Era of Tailor-made Jobs: Second Report on Local Development and Employment Initiatives*. SEC 98, 25 January. Brussels: European Commission.

CEC (1998b) *The Third System and Employment: A First Reflection*. Brussels: European Commission D5/A4 and the European Parliament.

CEC (2000) *Local Enterprising Localities: Area-based Employment Initiatives in the United Kingdom*. European Union Regional Policy Study 34 (DGXVI). Luxembourg: European Commission.

Dean, M. (1991) *The Constitution of Poverty: Towards a Genealogy of Liberal Governance*. London: Routledge.

DETR (1998) *New Deal for Communities*. London: Department of the Environment, Transport and the Regions.

Driver, F. (1993) *Power and Pauperism: The Workhouse System, 1834–1884*. Cambridge: Cambridge University Press.

DTZ Pieda Consulting (1999) *Review of Management, Structure and Role of Organisations and Projects in Craigmillar: Final Report*. Edinburgh: City of Edinburgh Council, 9 March.

Ekins, P. and Newby, L. (1998) 'Sustainable wealth creation at the local level in an age of globalisation', *Regional Studies*, 32 (9): 863–71.

Frazer, E. (1999) *The Problems of Communitarian Politics: Unity and Conflict*. Oxford: Oxford University Press.

Giddens, A. (1998) *The Third Way: The Renewal of Social Democracy*. Cambridge: Polity Press.

Gittell, R. and Vidal, A. (1998) *Community Organizing: Building Social Capital as a Development Strategy*. Thousand Oaks, CA: Sage.

Glennerster, H., Lupton, R., Noden, P. and Power, A. (1999) *Poverty, Social Exclusion and Neighbourhood: Studying the Area Bases of Social Exclusion*. CASE Papers No. 22, March. London: LSE Centre for the Analysis of Social Exclusion.

Greffe, X. (1998) 'Local job development and the Third System', paper presented to the *Third System and Employment Seminar*, held jointly by the European Parliament's Employment and Social Affairs Committee and DG5 of the European Commission, 24–25 September. Published in DG5 (1999) *Third System and Employment Seminar Proceedings* (CE-V/1-99-0004-EN-C). Brussels: CEC. Copies available from 200 rue de la Loi, 1049 Brussels.

Grimes, A. (1997) 'Tuning into the Third Sector', *New Economy*, 4 (4) (Winter): 226–9.

Haughton, G. (ed.) (1999) *Community Economic Development*. London: HMSO/Regional Studies Association.

Hayton, K. (2000) 'Scottish Community Business: an idea that has had its day?', *Policy and Politics*, 28 (2): 193–206.

Henderson, K.M. (1999) 'A Third Sector alternative: NGOs and grassroots initiatives', in K.M. Henderson and T. Dwivedi (eds), *Bureaucracy and the Alternatives in World Perspective*. London: Macmillan. pp. 52–68.

Hetherington, P. (2000) 'Out of stock', *The Guardian*, 6 September.

Jessop, B. (1993) 'Towards a Schumpeterian workfare state? Preliminary remarks on post-fordist political economy', *Studies in Political Economy*, 40 (Spring): 7–39.

Jessop, B. (1995) 'Towards a Schumpeterian workfare regime in Britain? Reflections on regulation, governance, and welfare state', *Environment and Planning A*, 27 (10): 1613–26.

Keynes, J.M. (1933) *The Means to Prosperity*. London: Macmillan.

Keynes, J.M. (1936) *The General Theory of Employment Interest and Money*. London: Macmillan.

Levitas, R. (1996) 'The concept of social exclusion and the new Durkheimian hegemony', *Critical Social Policy*, 16 (46): 5–20.

Levitas, R. (1998) *The Inclusive Society? Social Exclusion and New Labour*. London: Macmillan.

Marx, K. and Engels, F. ([1872] 1967) *The Communist Manifesto*. London: Penguin.

O'Connor, J. (1973) *The Fiscal Crisis of the State*. London: St Martin's Press.

Pearce, J. (1993) *At the Heart of the Community Economy: Community Enterprise in a Changing World*. London: Calouste Gulbenkian Foundation.

Polanyi, K. (1957) *The Great Transformation: The Political and Economic Origins of Our Time*. New York: Rinehart.

Power, A. and Bergin, E. (1999) *Neighbourhood Management*. CASE Papers No. 31, December. London: LSE Centre for the Analysis of Social Exclusion.

Procacci, G. (1978) 'Social economy and the government of poverty', *Ideology and Consciousness*, 4: 55–72.

Reisman, D. (1991) *Conservative Captalism: The Social Economy*. London: Palgrave.

Rose, N. (1998) 'The crisis of the 'social': beyond the social question', in S. Hänninen, (ed.), *Displacement of Social Policies*. Jyväskylä: SoPhi. pp. 54–87.

Saunders, R. (1997) *Resident Services Organisations*. London: Priority Estates Project.

SEL (2000) *New Directions: Sustaining London's Communities*. London: Social Enterprise London. (Available from www.sel.org.uk or from 1A Aberdeen Studios, 22–24 Highbury Grove, London N5 2EA).

SEU (1998) *Bringing Britain Together: A National Strategy for Neighbourhood Renewal*, CM 4045. London: HMSO.

Smith, G.R. (1999) *Area-based Initiatives: The Rationale and Options for Area Targeting*. CASE Papers No. 25, May. London: LSE Centre for the Analysis of Social Exclusion.

Turok, I. and Hopkins, N. (1997) *Picking Winners of Passing the Buck? Competition and Area Selection in Scotland's New Urban Policy*. Glasgow: Centre for Housing Research and Urban Studies, University of Glasgow.

West, A. (1999) 'Regeneration, community and the social economy', in G. Haughton, (ed.), *Community Economic Development*. London: HMSO/Regional Studies Association. pp. 23–9.

World Bank (1997) *World Development Report*. New York: Oxford University Press.

three

Alternative Financial Spaces

Duncan Fuller and Andrew E.G. Jonas

Although geographers have had a longstanding interest in finance and the spatial circulation of capital (e.g., Harvey, 1982), research in the 'new economic geography' has only recently begun to examine 'alternative' financial institutions and the spaces these institutions occupy (Lee, 1996, 1999). In the case of one such 'alternative' institution (Gunn and Gunn, 1991), the British community credit union, the local development space is usually referred to as the common bond area, which in this instance is delimited by the area of residence and/or workplace of members of the credit union.[1] The common bond area serves as a basis of mutuality for the credit union and provides the geographical boundaries within which the pooling of savings and lending of money at relatively low cost to members occurs. It is tempting to examine 'alternative' institutional forms like community credit unions solely in terms of social and economic conditions within their common bond area rather than, say, the wider (national and international) political and economic contexts. This is not to say that the national (or even the international) contexts are analytically *un*important. Indeed, the development of 'alternative' local spaces of finance has much – although not all (Ford and Rowlingson, 1996) – to do with the withdrawal from these spaces of 'mainstream' financial institutions in response to rapidly-changing national regulatory frameworks and global markets (Leyshon and Thrift, 1997). Rather, there is an assumption that the development of 'alternative' financial institutions is first and foremost dependent upon social relations and organisational capacities specific to a local area or community; for these provide a geographical basis for financial autonomy and resistance to wider economic trends (cf. Lee, 1999).

Such a view is inherent within the general direction of recent work in economic geography, which seeks to identify the diverse ways in which economic forms are culturally and geographically embedded and, in turn, how economies are actively constructed in place and space as, variously, networks, forms of association, geographical flows, and/or scalar hierarchies (see contributions in Lee and Wills, 1997). Beyond the issue of understanding how 'alternative' financial forms become embedded in the landscape, however, there is the important question of how these forms and their supporting social networks, flows and institutional hierarchies are (or are not) reproduced over time. For in providing individuals and communities with 'alternative' means of access to financial resources, small-scale institutions like community credit unions do not simply facilitate social reproduction in place but must also be reproducible as sustainable and 'locally alternative' financial institutions. Moreover, there is a danger of representing 'alternative economic spaces' in somewhat static or even idealist terms – as places of economic survival, social resistance, and financial autonomy – as compared to the 'dynamic' or 'real' spaces of monetary exchanges, capital flows, and, ultimately 'globalization'. For example, it has been argued that:

> the move towards local monies and local community currencies is motivated not only by the desire to bridge the quantitative gap between what individuals earn and what they need to survive financially and socially, although this may, of course, be an important factor. It is also a community-building tool, a process of resistance against unacceptable values, a means of revitalising depressed and impoverished local economies and communities which have become marginalized from the monetary and economic mainstream, and a way of constructing alternative economic geographies founded on different social relations and ... value[s]. (Lee, 1999: 210)

The vision of 'alternative' local economic spaces implied here is one that imbues these places with distinctive *social* values and ideals that stand in contrast to *economically* rational/normative values supposedly at large in the wider space economy. However, this begs the question of how these spaces and places in practice preserve such values and ideals, their local autonomy, and the social, economic and political bases of their 'alternative' nature.

In this chapter, we concentrate on two main issues. First, we examine recent changes within the British credit union movement in order to argue that local 'alternative economic spaces' are not impervious to wider economic and political developments. Secondly, and in so doing, we problematize the notion of 'alternative' institutions, and their role in creating and sustaining 'alternative' economic spaces. Here we propose a more nuanced distinction between 'alternative' institutions, and, by implication, the nature of the 'alternative' economic spaces they occupy, in three main ways. First, we suggest there are those that are engaged in the process of actively and consciously *being* alternative – as embodying something 'different' in value or operational terms, whilst simultaneously representing a rejection of more non-alternative, or 'mainstream' forms and their identities. These we term *alternative-oppositional* institutions. Secondly, we suggest there are 'alternative' institutions that represent an additional choice to other extant institutions, whilst not necessarily being distinguishable in the sense of *being* alternative described above – *alternative-additional* institutions. Thirdly, we suggest there are 'alternative' institutions that can act as a form of substitute (or even as institutions of last resort) for institutions that are no longer present, and which may or may not be engaged in *being* alternative (*alternative-substitute*). In these terms, the kinds of 'alternative' economic spaces described by Lee are perhaps more accurately definable as *alternative-oppositional economic spaces*, reflecting the embedded, conscious desire to reject other, largely mainstream-related values and ideals.

Using these ideas, the chapter proceeds as follows. We begin by highlighting that, in light of its unique development history, the community credit union in Britain in many respects represents an ideal example of a locally alternative-oppositional financial institution. However, we go on to critique this view, and to reflect upon attempts underway to change the notion of what a credit union is – and for whom it exists. In particular we examine the ways in which closer ties are being drawn between the credit union movement (or more specifically, certain sections of the movement), 'mainstream' financial institutions, and the Labour government. We then move to consider how, as a consequence of these trends, the philosophical basis and role of community credit unions is being debated within the movement, leading to polarised positions that echo Berthoud and Hinton's (1989) distinction between 'idealist' and 'instrumentalist' models of credit union development, and which can be re-interpreted within the more nuanced notions of 'alternativeness' outlined above. Finally, we analyse how a key ideological battleground is emerging in relation to the local implications of these models 'on the ground'; that is, to the future capacity of community credit unions to construct/reproduce alternative-oppositional economic spaces in Britain. Arenas of struggle include credit union distinctiveness, the common bond (see Ferguson and McKillop, 1997), and the implication of attempts to generate the 'mix' of membership deemed essential in creating a new breed of commercially successful and sustainable community credit unions. In the final section, we provide a few tentative suggestions regarding the consequences of these issues, both for the credit union movement in Britain, and for understanding more complex notions of their 'alternativeness'.

CHALLENGE AND CHANGE IN THE BRITISH CREDIT UNION MOVEMENT

Although the credit union movement began in Germany in the 1850s, it was not until the 1960s that the movement gained its first foothold in Britain (National Consumer Council, 1994). Table 3.1 shows the state of the movement in 1999, worldwide, across Europe, and in Britain. Although the data are not entirely accurate, under-representing the movement at all scales because they only include credit unions affiliated with World Council of Credit Union-affiliated organizations (WOCCU) such as the Association of British Credit Unions Limited, the table usefully demonstrates that Britain accounts for approximately one per cent of all credit unions world-wide, and much less than one per cent of total worldwide membership. In addition, less than one per cent of British adults are credit union members, as compared to membership rates of between thirty and fifty per cent in countries such as Ireland, the United States and Australia (National Consumer Council, 1994). In general terms, therefore, the movement is significantly under-represented in Britain.

Table 3.1 The British credit union movement in international context (all data are for 1999, $US millions)

	World-wide	Europe	Great Britain
Credit unions	37,759	5,798	468
Members	100,826,082	4,550,184	228,312
Savings	$407,391m	$5,036m	$210m
Loans	$314,219m	$2,912m	$211m
Reserves	$46,688m	$671m	$23m

Source: Adapted from World Council of Credit Unions, Inc. (http://www.woccu.org/)

The present structure and philosophy of the credit union movement in Britain can, by in large, be traced to the 1979 Credit Unions Act (see, for example, National Consumer Council, 1994: 5–7). This Act required all credit unions to be organized around a common bond, which serves as the basis of mutuality and trust for members of the credit union. Once the common bond has been defined, individual membership is determined not by creditworthiness or previous financial 'health', but by the ongoing ability and willingness to save; members' savings are pooled within the common bond. The majority of credit unions in Britain are common-bonded on the basis of community; that is, they operate as financial co-operatives, owned and run on a voluntary basis by their members, all of which reside and/or work in the common bond area. In this respect, the community common bond serves as a mechanism of accountability among the members, with the assumption that if members come from broadly the same area or neighbourhood they are more likely to know – and therefore to trust – each other. As we will argue later on in the chapter, this lends to the community credit union an idealist view of its underlying philosophy and organizational structure (see Berthoud and Hinton, 1989). This view, however, which is becoming increasingly untenable.

Since they first emerged in Britain in the 1960s, community credit unions have perhaps been regarded, implicitly at least, as quintessentially alternative-oppositional financial institutions. Initially set up by local church groups and volunteers from Afro-Caribbean communities in urban areas such as South London, the movement later spread across Britain. Throughout this period community credit unions have generally performed well in providing individuals and communities with *access* to credit and savings opportunities (Fuller and Jonas, 2002a, 2002b). At the same time, as the credit union movement has expanded into new areas, different community credit union development 'models' have flourished, often reflecting the heterogeneous social and economic conditions to be found in the localities and areas they inhabit (Fuller and Jonas, 2002a). Various national studies (e.g., National Consumer Council, 1994) have shown that the potential for the further growth and spread of credit unions across Britain is significant, and with it the possibility of social and financial empowerment for those individuals who become members of a community credit union. Credit unions are also recognized to be increasingly vital to the economic stability of localities and communities otherwise adversely affected by economic restructuring and financial withdrawal (Conaty and Mayo, 1997; Fitchew, 1998). In summary, credit unions have rapidly moved from the status of locally alternative-oppositional financial institutions to a position in which their potential is increasingly viewed as an important facet in contributing to the construction of a new, socially inclusive British financial landscape, as alternative-additional/substitute forms.

Viewed in this context, the community credit union would seem to be an excellent example of an institution potentially capable of creating, sustaining, and representing alternative–oppositional economic spaces. However, and as is implied above, over the next few years the local development context will be increasingly influenced by national-level regulatory and institutional changes, largely as a consequence of the perceived uneven development of community credit unions, the difficulties of sustaining such credit unions in impoverished areas, and the potentially disastrous impacts such 'failures' could have on their suggested roles in enhancing financial and social inclusion more generally. These changes, in turn, present a number of challenges for the credit union movement and local credit union organizations, including: attempts to redefine the 'model' credit union within the national credit union movement; the changing regulatory context for credit union development and New Labour's attempts to embrace credit unions within policies on social exclusion; and what we have elsewhere defined as the 'local' challenge – the way in which these developments impact 'on the ground' in a variety of local credit union development contexts (Fuller and Jonas, 2002a, 2002b). Of particular significance is the Labour Government's growing support for credit union development, which is seen as central to its policies on social and financial exclusion (Social Exclusion Unit, 1998; HM Treasury, 1998a, 1998b, 1999a, 1999b, 1999c; HM Government, 1999). In recognizing the growing interest on the part of New Labour in what credit unions are or should be, representatives of the wider credit union movement have called for credit unions to adopt a 'new direction' (HM Treasury, 1999c: 9). This call in turn relates to the ways in which understandings of credit unions have been constructed and contested in Britain since the earliest days of the movement.

Berthoud and Hinton (1989) argued that British credit union development was comprised of two contrasting ideological perspectives, which continue to be of relevance today. On the one hand, those from a social or *idealist* approach advocate that a main objective of credit union development is to help people on low incomes, especially if they are excluded from access to more 'mainstream' financial sources. In addition, an emphasis is placed on the participation of members in setting up and running their own institution as a form of empowerment. Such credit unions:

> need to be kept small, to ensure that individual members enjoy genuine participation in the management of their own group. The common bond should be based on poor communities, rather than on existing institutions whose members have money to save. Caution is needed to prevent more prosperous members from 'highjacking' the management and/or diverting the union's activities to meet their own requirements. Loans should be made to people who need credit, rather than to those in the best position to repay the money. (Berthoud and Hinton, 1989: 7)

Emphasis here has been placed on the role of the credit union in facilitating a process of community organization and empowerment, with a view to creating more sustainable communities and local economies. Some might argue that such a model of credit union development is broadly consistent with contemporary theories of social capital formation and democratic participation as found in, for example, recent critiques of mainstream institutional economics (Gunn and Gunn, 1991; Putnam, 1993). On the other hand, and in contrast to this first model, there is a more *instrumentalist* model of credit unions, which sees them as financial institutions that provide 'a medium of exchange between savings and loans [as] an end in itself' (Berthoud and Hinton, 1989: 7). Here 'management objectives of efficiency and financial stability take priority over considerations of procedure and participation' (Berthoud and Hinton, 1989: 7). Such services are suggested to be as 'useful' to those with more money as they are to those on low incomes, with a preference for including as many people as possible, most notably through an emphasis on larger common bonds than those espoused from an idealist perspective. The reality of local development tends to suggest that these different models should be viewed as extremes on a continuum of ideological perspectives rather than an either/or dualism. Nonetheless, in the British experience credit union development – particularly local development – has tended to be informed by the social or *idealist* model. Perceived as the 'poor persons bank', espoused as a panacea for community/community economic/local economic development, and adopted as a core feature of many anti-poverty initiatives, the roles of (especially community) credit unions in the past have been generally narrowly defined, serving the financial needs of (often, *the* most) impoverished individuals and communities.

Recently, however, critics from the instrumentalist persuasion have argued that many credit unions have failed in generating the income/financial surpluses necessary to make a fundamental impact on the financial circumstances of their members (Jones, 1998, 1999). They have argued that organizational and financial problems have led to difficulties for community credit unions in developing into locally sustainable and effective alternative sources of credit and finance for *all* potential members within these areas. Prior to the last few years such debates and disagreements were voiced within the confines of the movement. However, in recent years such debates have entered into a broader public/political forum (see for example, Toynbee, 1999). By the same token, it has long been acknowledged that differences in development 'success' exist between community credit unions and their associational and occupational counterparts. This relates to the manner in which these latter forms can take advantage of pre-existing membership fields, operating systems, routes of information and publicity, and so on. There has been a tacit acceptance that such credit unions will, in all likelihood, always be more 'successful' in terms of their asset-member ratios. However, recent arguments suggest that such views have overlooked a number of underlying weaknesses within the community credit union model, evidenced by a gradual reduction in the asset base of community-based credit unions over time, especially in relation to the other types of credit union. Commentators representing the more instrumentalist section of the British movement (such as the main national credit union trade association, the Association of British Credit Unions Limited, ABCUL) have argued that research demonstrates that the lack of success is a consequence of community credit unions operating within an essentially 'flawed', 'old model' of credit union development.

This 'old model' is based upon 'an interrelated network of assumptions, beliefs, understandings and commitments about credit union development that has produced, in the minds of communities, local authorities and volunteers, a certain model of credit union organization, purpose and structure' (Jones, 1998: 5). The 'old model' characteristics of 'community-based' credit unions are: their small size; their failure to generate sufficient income or surpluses to achieve financial self-sufficiency and stability; that many verge on insolvency; that they have overworked, stressed and ageing volunteers; and, that consequently, they serve only 1–2 per cent of their potential membership fields. These same commentators have subsequently voiced concerns regarding both the sustainability of community credit unions (in terms of volunteer energy, entrepreneurialism, skills and so on), and the potential disastrous impact of significant numbers of poorly performing credit unions for the movement as a whole. As such they have argued for the need to move towards a 'new model' of credit union development (Jones, 1998) which has more than a passing resemblance to the instrumentalist vision of credit union development. This 'new model' has the aim of being commercially successful and socially inclusive, in addition to retaining credit union mutuality, and will involve: the redefinition of the concept of 'small' within the credit union context (that is, the development of larger common bond applications, and consolidation of smaller credit unions through mergers, take-overs, and closures); a reworking of the rôles of credit union volunteers; an understanding of how to run a fully-professional financial service; and the introduction of paid staff to carry out day-to-day activities.

At the same time, as we have already noted, credit unions are gradually being absorbed into the mainstream of social and economic policy-making in Britain. The Labour Government has been keen to highlight the positive impacts credit unions can make in terms of combating social exclusion and financial withdrawal (Fitchew, 1998; Social Exclusion Unit, 1998). In these terms, as part of six initiatives designed to 'help people in disadvantaged communities excluded from key financial services' (HM Treasury, 1999a: 1), a strong emphasis has been placed on the need to 'boost' credit union development within Britain. This is to be achieved through changes to the overly restrictive regulatory framework for credit unions (Fitchew, 1998), and the creation of a new Central Services Organisation (CSO) to support and enhance the role of credit unions in the future. Clearly, these developments should be viewed in the context of the selective dismantling of the welfare state, with greater responsibility for social reproduction placed on individuals, communities, and localities across Britain (see Amin et al., 1999).

In sum, these trends appear to signal the appropriation of the role, identity and philosophy of British credit union development (in its newest guise) by the state (albeit with help from certain key players from within the credit union movement itself), and given legitimacy through social exclusion policy. This appropriation has clear implications for the future development trends, development trajectories, and alternative-oppositional nature of local British credit unions, alongside their role in creating 'alternative-oppositional economic spaces'.

COMMUNITY CREDIT UNIONS AS ALTERNATIVE-OPPOSITIONAL FINANCIAL INSTITUTIONS

We have suggested elsewhere (Fuller and Jonas, 2002a) that perhaps one consequence of the youthful focus within geographical research on money and finance (see Martin, 1999) is the continuing quandary surrounding the description and categorization of the many different institutional forms that comprise any financial services 'industry'. Seemingly interchangeable and binary oppositions have been utilized in order to express these differences, such as notions of an institution's formality (formal/informal), aspects of their regulation (regulated/unregulated), their visibility (high street based/network based), their motivations (economic outlook/social outlook), and position within the economy (mainstream/non-mainstream) amongst others. Recently, studies of the social relations and geographical contexts of *local* financial institutions and monies have tended to emphasize the 'alternative' nature of a number of these forms (for example see Lee, 1999), with the implication that these 'alternatives' are peripheral to some non-alternative 'core' essentially comprised of the main banks and building societies ('*the* mainstream').

However, at a time when credit unions are being appropriated by the state under a selective instrumentalist guise, their 'alternative' status, and categorization as such, is increasingly problematical. Two key reasons for this developing complexity relate to the development of the Central Services Organisation (CSO), and the proposed changes to credit union legislation. Although it has been argued that credit unions generally prefer support services (in relation to book-keeping, equipment, grants and so on for example) that are provided locally (National Consumer Council, 1994), at the time of writing there is a move to centralize such services in Britain through the creation of a CSO. The idea of a CSO comes from the (somewhat tautological) reasoning that 'strong' and 'successful' national credit union movements overseas are frequently associated with the presence of CSOs (HM Treasury, 1999c). It has been suggested that a British equivalent could provide a range of beneficial services to credit unions, either directly or through out-sourcing of key services to banks and other organizations with the appropriate skills and expertise. CSO services might include: back-office processing services to relieve volunteers of book-keeping and other 'administrative' tasks such as bill payments and other transaction services; assistance with business planning and financial management; assistance with member financial education and marketing; the provision of a treasury management facility; assistance with product development; recycling surpluses from credit unions with an excess of savings to those with an excess of borrowers; and general and encouragement and support at each development stage (HM Treasury, 1999c; Social Exclusion Unit, 1999). However, the creation of a CSO opens up opportunities for banks and building societies to develop closer ties to credit unions throughout Britain, thereby blurring the simplistic distinction between credit unions as 'alternative' financial institutions, on the one hand, and the financial 'mainstream', on the other (Fuller and Jonas, 2002a). While a CSO structure (and its operating principles) has yet to be finalized, ABCUL is forcefully representing itself as the key organization capable of co-ordinating its development and operation. A regional structure has been rejected on the grounds that economies of scale would be reduced; in addition, the extent to which 'standardized' services would allow for local contextual differences and requirements to be incorporated in the delivery of these services is uncertain. Clearly therefore, it can be argued that support service developments are explicitly striving to make the national credit union model *per se* less alternative-oppositional in nature.

Proposed deregulatory measures also have large implications for the degree to which credit unions can or will remain distinct from other financial institutions. Here, the proposed regulatory reforms include: increases in the maximum repayment period for loans; greater flexibility in the common bond requirements; alignment of the maximum amounts that can be held in youth accounts with adult account levels; the removal of the maximum membership limit for individual credit unions; allowing credit unions to charge for ancillary services; greater flexibility on the disposal of re-possessed collateral; further consultation on increasing the sources from which credit unions can obtain credit; and greater flexibility on dividend accounts (HM Treasury, 1999b). These proposals come at a time when the regulation of credit unions is passing from the Registry of Friendly Societies to the Financial Services Authority, with the message that 'stronger regulation is the natural counterpart of a stronger credit union movement' (Social Exclusion Unit, 1999: 17).

As such, community credit unions are currently operating in a rapidly changing national environment, the effects of which will have important implications for both how credit unions are perceived in the wider British society, and their position within the financial services industry as a whole. Clearly, many of the details are yet to become clear. However, a number of potential scenarios can be suggested. First, depending on the 'success' of the 'new model', credit unions could become a true source of competition to the financial 'mainstream' (as alternative-additional forms) as they become increasingly similar in terms of the facilities they can provide, their visible presence, their impact, and overall identity. In so doing, key questions remain concerning how credit unions retain their ideological and philosophical distinctiveness, especially in relation to those whose perception of what credit unions are lies firmly in the social, idealist, or alternative-oppositional camp, and/or what degree of support might be forthcoming from the 'mainstream' players (as potential competitors). Second, there is the distinct possibility that credit unions are being groomed into a form of second-tier banking service, effectively as alternative-substitutes. This might allow other ('mainstream'?) financial players to continue to target the wealthier sections of society, whilst relieving their conscience (and bad publicity) through participation in the development of an apparent 'alternative' for those left behind, albeit one that seems increasingly removed from the alternative-oppositional forms highlighted in recent academic work on local monies. This direction would be consistent with the trend towards the dismantling of the welfare state as increasing emphasis is placed on individual responsibility, alongside selective attempts to foster the development of the 'social economy' across parts of Britain.

These different scenarios become more complicated at a local level. Recent regulatory proposals suggest a distinction between two main forms of credit union development. On the one hand, there are those who can demonstrate the capability of ensuring the adequate delivery of an enhanced range of financial products and services than is available at present (presumably those who will follow a 'new model' perspective). On the other hand, there are those who either do not wish to take such an instrumentalist line or cannot demonstrate sufficient financial and organizational 'health' to allow the move to such status to be reached. Proponents of the idealist model – who largely reject instrumentalist proposals – may even be forced to deregister institutions as credit unions, and re-register them as savings clubs. Presumably this would protect the brand image of the 'new' variants, and remove the possibility of a variety of different style credit union forms between localities.

Clearly many questions remain unanswered concerning the impact of these proposals and developments on the identity and status of community credit unions in Britain. However, it is clear that the traditional depiction of them as inherently different and/or 'alternative' to the rest of the financial service industry is increasingly problematic, unstable, and unpredictable at both national and local scales where the diversity of development styles and principles muddies the water further. More generally, it can be confirmed that such binary categorizations and typologies may in fact be misleading rather than insightful, and increasingly unsupportable in a 'Third Way' Britain in which the distinctions between work and welfare, 'alternative' or 'mainstream', etc., are becoming increasingly blurred, if not contested (but see Amin et al., 1999).

COMMUNITY CREDIT UNIONS AS *LOCALLY* ALTERNATIVE-OPPOSITIONAL INSTITUTIONS

Many of the developments outlined above appear to move (or seek to move) credit unions – especially community credit unions – away from their more local, social, idealist, alternative–oppositional roots, and towards a more national, economic and instrumentalist development structure. 'Success' is increasingly being framed and scrutinized in terms of key, measurable 'economic' impacts, such as the proportion of members in relation to the total possible membership base, the amounts they can and will borrow, the range of financial products that can be offered to this population, and the overall asset base and member/asset ratios of credit unions. In addition, it appears that the main proponent of the 'new model', ABCUL, is increasingly aligning itself with the international credit union movement, in the process representing itself as *the* organization representing the movement in Britain. This raises important questions in relation to those credit unions and sections of the movement whose identity is based around a more local, social idealist focus (with harder to articulate, social impacts and goals), and whose main reason for existing can be cited more in terms of alternative-oppositional notions of promoting community cohesion and local financial support structures than in creating new financial institutions for the wider British economy.

Here, as noted above, the movement in Britain has been built on more idealistic, if at the same time practical, *local* alternative oppositional foundations. In many ways local credit union development contexts in Britain are similar to how Filion (1998) characterizes community economic development (CED), as 'imbued with a deep attachment to the community and a powerful will to survive in the face of economic adversity' (Filion, 1998: 1115). Similarly, we could draw links between how credit union development in many localities has occurred and Haughton's (1999) characterization of CED as a process of supporting local economic strategies that help to 'restructure community' rather than 'restructure capital'. In these terms, it is unclear how wider economic and political developments will impact on local contexts where credit unions have been espoused as focusing as much on social issues and community development, than the specifically economic. There are many in the movement who believe that, in taking steps towards this more instrumentalist, economically oriented outlook, the basic notion of what a credit union is in itself is being explicitly eroded. The National Association of Credit Union Workers (NACUW) has suggested that 'it would be a great failure if credit unions became so obsessed with and focused on their economic purpose, *that they became just another financial institution*' (NACUW 1999: 2, *emphasis added*). This statement reflects a concern both over the impact on the apparent alternative-oppositional nature of credit unions, and the potential re-direction of their roles and identity.

Certainly at the local level, the roles of credit unions continue to be diverse and locally specific. In many areas local authority support has been crucial for setting up new credit unions (Thomas and Balloch, 1992), or ensuring the survival of 'failing' established credit unions. In addition, local credit unions workers often have different development views and ideologies. Many workers believe that credit union development should remain first and foremost a voluntary, community-based activity, led from the grassroots with development roles and priorities being defined locally, by local 'experts', and with minimal public assistance. By contrast, civil servants and even local authority officers are often seen to hold an instrumentalist view of credit unions, which encourages them to see credit unions simply as components of wider anti-poverty or regeneration strategies.

A further difficulty relates to public perceptions of credit unions. Frequently, one reads dismissive references to the 'lack of interest in' or 'ignorance of' credit unions among the British public. This dismissive attitude may have been reinforced by research that claims that many credit unions – especially local authority-assisted community credit unions – are close to 'failure' and cannot survive in their current form as stand-alone 'alternative' providers of credit (Jones, 1998, 1999; and see Toynbee, 1999). It may also reflect the proliferation of new forms of 'mainstream' financial institutions and public exposure to them through the media and the Internet. Although this same research recognizes that there is considerable potential for the further development of credit unions, there is growing uncertainty about the form of this development. Although space remains at the local level for community credit unions to survive and expand in their role as alternative-oppositional financial institutions, the local development context is becoming increasingly contested as a more instrumentalist view of credit union development takes root.

CREDIT UNIONS AND ALTERNATIVE-OPPOSITIONAL ECONOMIC SPACES: COMMON BONDS, SPACE, PLACE, AND SCALE

It is expected that, if adopted, the regulatory and organizational proposals outlined earlier, combined with the move to a 'new model' approach, would result in many of the smaller or 'failing' credit unions being merged with, or taken over by, larger and more commercial credit unions. Such issues have important implications concerning how the 'new economic geography' is to study credit unions, particularly in relation to the potential impacts the challenges and developments may have on the defining feature of credit union development, the common bond (Ferguson and McKillop, 1997). Essentially, understanding the development of the common bond for community credit unions – the association or link between members – takes us beyond static concepts of place into more dynamic, social-constructivist ideas of spatiality and the politics of scale (see, Smith, 1992; Jonas, 1994; Marston, 2000; Brenner, 2001). By its very nature, the common bond represents the confluence of the cultural, the economic and the spatial domains; it is this *socially and spatially simultaneous form of association* that binds together the membership as a collective interested in social reproduction through the exchange of money and credit. Moreover, it has traditionally been articulated as embodying the 'place' or 'locale' around which the credit union is constructed, thereby imbuing each credit union with its own sense of mutuality, uniqueness, and identity, whilst also serving as a fundamental controlling element on the behaviour of credit union members.

Yet in the face of the challenges confronting the movement in the years to come, there will be important implications for the role of the common bond in future development. Here tensions between place and space, between jumping geographical scales (Smith, 1992) and/or constructing new social 'networks' and capital 'flows' across space (Brenner, 2001), will become increasingly important, especially as moves are made away from small, 'local' or 'community' defined common bond areas, to larger common bonds and new organizational structures (such as a national CSO). Also tensions will occur as credit unions struggle to incorporate the 'mix' of membership that is thought essential to the success of sustainable credit union development. In the British context, this may require a move from common bonds based around housing estates, or apparently distinct local area 'communities' where members 'know each other', and where internal policing and the power of shame is brought to bear in local networks, to more anonymous city-wide common bonds or to structures that rely on spatially extended networks of association rather than those constructed around 'local territories'. In these circumstances, questions may arise both in relation to the identity of the credit union and to how its members relate spatially and by association to the larger institution (see, for example, Fuller, 1998). Thus there are important strategic and political issues concerning how this 'jump' in scale (or extended networking) is managed. These include convincing credit unions that the 'jump' can, or even should, be made in the first place, while at the same time maintaining what many people see as a core feature of credit union development – its distinctive ability to respond to the needs and requirements of 'local' people.

CONCLUSION

In Britain, it can be suggested that community credit unions have traditionally played a role in carving out alternative-oppositional economic spaces for individuals and communities seeking access to convenient, low-cost ways of saving and borrowing (McArthur et al., 1993; National Consumer Council, 1994). However, as banks and building societies continue to restructure nationally and withdraw from localities (Leyshon and Thrift, 1997), moves are afoot to enhance the extent to which credit unions – particularly community credit unions – can offer a more stable financial and institutional presence for people in those localities, especially for those on low incomes and denied access to cheap credit. Struggles are occurring around and within the national development space for credit unions, struggles into which local credit union organizers are being drawn, willing or otherwise. If a new instrumentalist credit union development model is adopted, it would help to establish a more legitimate national presence for credit unions as alternative-additional/substitute forms (and representatives of the movement) in Britain, thus helping the movement to grow further. In so doing the space for community credit unions to retain their financial autonomy and alternative-oppositional identity may be contracting. At the local level however, it is still possible to identify distinctive credit union development 'models', and it is these that continue to underpin the creation of alternative-oppositional economic spaces of credit union development. The national focus on the 'economic' side of credit union roles and operations belies a range of more socially-oriented alternative-oppositional development ideologies and philosophical approaches at the local level which, when interpreted in conjunction with the spaces credit unions inhabit, represent key battlegrounds in resisting the more standardized, national proposals (Fuller and Jonas, 2002a, 2002b).

An underlying assumption of research to date on 'alternative economic spaces' is that financial institutions occupying these spaces must retain their local distinctiveness, which is the basis of their autonomy from, and resistance to, the financial 'mainstream'. The process of creating these spaces is a means of simultaneously building community solidarity, resisting unacceptable values, and revitalizing depressed and impoverished local economies and areas (Lee, 1999). However, we have argued that the reproduction of such spaces as representing 'alternatives' is problematic; the idealist vision necessitated by seeking to protect these spaces from unwarranted external forces neglects a reality of struggle both within and between 'alternative' economic institutions and their spaces. For researchers interested in understanding the future role of community credit unions in the construction of 'alternative' economic spaces, the challenge is to approach these issues in a more nuanced manner; to identify which credit union development models manage to resist the national trend and sustain some degree of organizational, institutional, alternative-oppositional autonomy from centralization tendencies within the movement. Community credit unions may well continue to play an important role in defining alternative-oppositional institutional pathways and spaces in Britain; but it is a role that is increasingly shaped by forces emanating beyond any given local development context, and the pressures for change are increasing.

Note

1 This chapter examines community credit unions, which comprise a majority of credit unions in Britain. Britain also has examples of associational and occupational credit unions, including some that are operated by local authorities for the benefit of their employees.

References

Amin, A., Cameron, A. and Hudson, R. (1999) 'Welfare as work? The potential of the UK social economy', *Environment and Planning A*, 31, 11, 2033–51.

Berthoud, R. and Hinton, T. (1989) *Credit Unions in the United Kingdom*. London: Policy Studies Institute.

Brenner, N. (2001) 'The limits to scale? Methodological reflections on scalar structuration', *Progress in Human Geography* 25(4): 591–614.

Conaty, P. and Mayo, E. (1997) *A Commitment to People and Place: The Case for Community Development Credit Unions*. Report for the National Consumer Council. London: New Economics Foundation.

Ferguson, C. and McKillop, D. (1997) *The Strategic Development of Credit Unions*. Chichester: Wiley.

Filion, P. (1998) 'Potential and limitations of community economic development: individual initiative and collective action in a post-Fordist context', *Environment and Planning A*, 30: 1101–23.

Fitchew, G. (1998) Address by the Chief Registrar of Friendly Societies to the World Council of Credit Unions conference, Glasgow, 26 May.

Ford, J. and Rowlingson, K. (1996) 'Low-income households and credit: exclusion, preference, and inclusion', *Environment and Planning A*, 28: 1345–60.

Fuller, D. (1998) 'Credit union development: financial inclusion and exclusion', *Geoforum*, 29 (2): 145–58.

Fuller, D. and Jonas, A.E.G. (2002a) 'Institutionalising future geographies of financial inclusion: Challenges and changes confronting the British credit union movement', *Antipode,* 34 (1): 85–110.

Fuller, D. and Jonas, A.E.G. (2002b) 'Capacity-building and British credit union development', Local Economy, 17 (2): 157–63.

Gunn, C. and Gunn, H.D. (1991) *Reclaiming Capital*. Cornell University Press. London.

Harvey, D. (1982) *The Limits to Capital*. Oxford: Blackwell.

Haughton, G. (1999) 'Community economic development: challenges of theory, method and practice', in G. Haughton (ed.) *Community Economic Development*. Norwich: The Stationary Office. pp. 3–22.

Her Majesty's Government (1999) *A Better Quality of Life: A Strategy for Sustainable Development for the United Kingdom*. London: HMSO.

HM Treasury (1998a) Deregulation and contracting out Act 1994: proposed amendments to the Credit Unions Act 1979, Regulatory reform. Consultation document. London: HMSO.

HM Treasury (1998b) 'More credit for the credit unions'. HM Treasury News Release, 192/98, 16 November.

HM Treasury (1999a) 'Initiatives to tackle financial exclusion'. HM Treasury News Release, 190/99, 16 November.

HM Treasury (1999b) 'Enhanced role for credit unions'. HM Treasury News Release, 191/99, 16 November.

HM Treasury (1999c) *Credit Unions of the Future*. Report of the Credit Union Taskforce, November.

IALLG (Inter-Association Legislative Liaison Group) (1997) Common ground: national goals for improving the laws governing credit unions: a report from the credit union movement of Great Britain. Unpublished MS.

Illsley, B. and Jackson, T. (1999) 'The third sector and local economic development in a peripheral Scottish city: the case of Dundee', in G. Haughton (ed.) *Community Economic Development*. Norwich: The Stationary Office. pp. 149–62.

Jonas, A.E.G. (1994) 'The scale politics of spatiality', *Environment and Planning D: Society and Space,* 12 (3): 257–64.

Jones, P.A. (1998) 'Towards Sustainable Credit Union Development: A Research Project. Association of British Credit Unions Limited, Manchester. Unpublished draft.

Jones, P.A. (1999) *Towards Sustainable Credit Union Development: A Research Project*. Association of British Credit Unions Limited, Manchester.

Lee, R. (1996) 'Moral money? LETS and the social construction of economic geographies in South East England', *Environment and Planning A,* 28: 1377–94.

Lee, R. (1999) 'Local money: geographies of autonomy and resistance?', in R. Martin (ed.) *Money and the space economy,* 207–24. Chichester: Wiley.

Lee, R. and Wills, J. (eds) (1997) *Geographies of Economies*. London: Arnold.

Leyshon, A. and Thrift, N. (1997) *Money/Space: Geographies of Monetary Transformation*. London: Routledge.
Marston, S. (2000) 'The social construction of scale', *Progress in Human Geography* 24(2): 219–42.
Martin, R. (1999) *Money and the Space Economy*. Chichester: Wiley.
McArthur, A., McGregor, A. and Stewart, R. (1993) 'Credit unions and low income communities', *Urban Studies*, 30(2): 399–416.
NACUW (National Association of Credit Union Workers) (1999) Towards sustainable credit union development: a response by the National Association of Credit Union Workers. Unpublished MS.
National Consumer Council (1994) *Saving for Credit: The Future for Credit Unions in Britain*. London: NCC Publications.
Putnam, R. (1993) *Making Democracy Work: Civic Traditions in Modern Italy*. Princeton, NJ: Princeton University Press.
Smith, N. (1992) 'Geography, difference and the politics of scale', in J. Doherty, E. Graham and M. Malek (eds) *Postmodernism and the Social Sciences*. New York: St. Martin's. pp. 57–79.
Social Exclusion Unit (1998) *Bringing Britain Together: A National Strategy for Neighbourhood Renewal*. London: HMSO.
Social Exclusion Unit (1999) *Access to Financial Services*. Report of Policy Action Team 14.
Thomas, I.C. and Balloch, S. (1992) 'Credit unions and local government: the role of metropolitan authorities', *Local Government Studies*, 18(2): 98–119.
Toynbee, P. (1999) 'The new credit deal for the poor is turning into an expensive disaster', *The Guardian*, 8 February.

four

Alternative Retail Spaces

Louise Crewe, Nicky Gregson and Kate Brooks

One of the ways in which we might begin to theorize 'the alternative' is in respect of labour processes and the spaces in which work takes place. As other contributors have noted, formal labour markets are fragmenting and becoming more differentiated, which is in turn raising conceptual questions about the meanings of work. Both the spatialities and the temporalities of work are said to be shifting, away from predictable and routinized working hours in large firms, towards much more fragmented, unpredictable and unstable working practices. And what is becoming clear is that analytical oppositions between formal/informal, paid/unpaid, mainstream/alternative work are becoming less and less useful in theorizing emergent labour processes within contemporary western economies.

In this chapter we address such shifting practices and spaces of work, looking particularly at work in the creative industries which have been seen as both emblematic of and in the vanguard of such shifts. Used as a shorthand to describe the convergence between the cultural and arts sectors and the media and information industries, the creative industries have become a potent symbolic territory in the emerging twenty-first-century knowledge economy. The sector includes those activities that have their origin in individual creativity and skill and that have a potential for wealth and job creation (Creative Industries Task Force, 1998). Including such sectors as advertising, design, fashion, film, music, software and multi-media technologies, the creative industries are argued to comprise some of the fastest-growing sectors in the economy and to account for a growing share of employment and output (Creative Industries Task Force, 1998). Their economic significance is underscored by raw quantifications which hint at exponential growth potential and powerful multipliers and spin-offs.[1] Such new independents are, it is argued, a driving force for growth and offer a new model of creative production.[2]

Yet while the cultural industries are increasingly being seen as alternative sources of employment within de-industrializing cities and while encouraging the creative city is beginning to bring together cultural and economic policy at an ever more strategic level (Landry, 2000; O'Connor, 1998; Pratt, 1997; Scott, 1999), little is actually known about how such sectors operate and how they can best be supported.[3] In addition, while there is now an increased sensitivity as to how issues of embeddedness and local networks help to create and sustain cultural milieux (Crewe, 1996; Crewe and Beaverstock, 1998; Grabher, 1993; Granovetter, 1991, 1992a, 1992b; Scott, 1996, 1999), oppositional polarities tend to dominate our conceptual grasp on the workings of such micro and small creative businesses. Those accounts which do exist tend to glamorize and glorify the creative dimensions of work in the cultural industries. This we would argue results from a failure to engage with the more mundane and ordinary aspects of work and to address the varied ways in which work gets done in the creative industries (Garnham, 1990). Attention has typically focused on the creative and aesthetic qualities of the sector rather than looking specifically at the ways in which labour processes are organized at sites involved in the production and consumption of symbolic commodities. So that while the cultural industries have been held up as symbolically significant, little attention has been paid to working practices in which creativity is the enterprise. As a direct result of this failure to address the varied ways in which work gets done in the creative industries, existing accounts have tended to either eulogize or dismiss the sector, oversimplifying complex realities and producing decontextualized accounts. This in turn has resulted in a tendency within current literature to present the cultural industries in terms of binary oppositions – as being either empowering, flexible, creative and fun, or, following the classic small-firm model, as risky, precarious, transitory and economically marginal. Such binaries in turn feed into broader political and policy debates about appropriate forms of work and about the nature and pre-conditions of economic competitiveness in the twenty-first century. All too often debates, particularly by those on the political left, draw unquestioningly on rhetorical overtures towards the knowledge economy, and look to a future dominated by small, creative firms, the new freelancers whose creative capacities are their competitive strengths (see DTI, 1998; Leadbeater, 1999). Within such work there has been, we argue, a reluctance to criticize the creative industries. Yet such uncritical acceptance of the creative work model is, we argue, highly questionable in the absence of any convincing evidence as to its economic, social and cultural contribution. So in terms of both conceptual accuracy and policy relevance, it is clearly important that we begin to move away from binary oppositions and polarities and to problematize the terms of the debate. This oppositional way of thinking, this either/or mentality, actually serves to lock us into debates about defining the boundaries between creative and conventional work/space, between the alternative and the mainstream, between the inconspicuous but interesting fringes and the ubiquitous but boringly self-evident mainstream. To conceptualize an alternative creative economy as an immutable, unmoving, definable fixed set of places or practices is, ultimately, largely unhelpful. And so what we go on to do in the following chapter is to interrogate critically the creative work model and to question the extent to which the cultural industries can be said to inhabit 'alternative spaces' and work in non-conventional ways.

As a means of opening up such debates we focus here on the working practices of one sub-group of symbolic producers, retro retailers and traders.[4] Retailers and traders, as we have discussed elsewhere, occupy a significant position in relation to the cultural industries more broadly (see Crewe et al., 2003; Gregson and Crewe, 2003). As well as being hitherto unresearched (and therefore of empirical interest), retro traders and retailers are also conceptually significant in that they are extremely ambiguously placed with respect to the creative industries more broadly. This ambiguity, we argue, forms the cornerstone to our understanding of their working practices and, in turn, sheds light on the potential ambivalences and tensions within the working practices of the cultural industries more generally. This ambiguity, as we go on to show, relates to retro retailers' and traders' limited scope for creativity. Unlike other cultural entrepreneurs, such as designers, artists and musicians, whose creativity is (largely) self-authored and self-produced, and whose businesses are built on the commercial application of creativity (Leadbeater and Oakley, 1999: 15), retro retailers' and traders' work rests on the (re)selling of previously commodified goods. Moreover, such commodities were typically located within mainstream fashion during their first cycle of consumption, and thus were neither particularly radical nor rare. In order to understand the craft involved in such enterprises we need to look not towards creative and artistic invention in its narrowest sense, but towards broader notions of tacit and situated knowledge and how this is mobilized through networks, alliances and embedded systems of social interaction. As we go on to show, knowledges and tastes in seeking out, reappropriating and recommodifying outdated fashions provides the basis for the competitive edge of retro exchange, rather than the artistic or creative talent of traders and retailers *per se*. Thus it is their cultural capital that provides the scope and site for their creativity and not, as with most other people who work in the creative industries, their artistic capacities expressed in self-authored commodities. This in turn suggests that such workers are much closer to, and more firmly embedded within, conventional mainstream retailing circuits than might at first appear. This indeterminate position, between the 'alternative' and 'the mainstream', we believe, may offer some important insights into the (re)creation and destruction of unconventional consumption spaces through unorthodox work practices.

For the purposes of this chapter, retailers are defined as concerned with sourcing, displaying and selling retro fashions and interiors through small (usually) independent shops, while traders operate on a more casual basis on stalls in a market setting such as Portobello in London, through small units such as those at Portobello Green, or in a warehouse setting such as Affleck's Palace in Manchester and Baklash in Nottingham. The geographies of retro trading and retailing offer some important insights into the spatialities of 'alternative' economies – both traders and retailers are typically located in definable cultural quarters within large university towns and cities such as Nottingham's Lace Market, Sheffield's Devonshire Road and Bristol's Park Street. The important point here, however, and the reasoning behind our making distinctions between traders and retailers, is that these 'alternative' spaces are not static and bounded but are constantly evolving in terms of the products which are displayed and the practices by which they are sourced, priced and sold. Subsequently, as we go on to discuss, the self-styled 'alternative' quarters of shops and marketplaces are thus deeply unstable and are constantly under threat from encroachment by more 'mainstream' concerns through ongoing commercial processes of property (re)valorization and gentrification, and through cultural processes of (re)commodification and shifting consumption imperatives. Political endeavours to conventionalize and normalize creative work practices further problematize the cultural and economic position of the creative industries and misunderstand the nature of their competitive success. Transformation and adaptation to an unpredictable and fickle marketplace are, we argue, inevitable defining features of the retro trade and any attempts to pin down and categorize the sector threatens to destroy its meaning and power. This innovativeness and preparedness to seek out change proactively is one of the means by which retro traders and retailers variously mark out their working practices and spaces as distinctive from, and alternative to, an imagined mainstream. In order to try to understand retro traders' and retailers' indeterminate and unstable positioning with respect to 'conventional' creative work practices and spaces, we adopt a methodology based on three different ways of reading retro landscapes. First, we analyse traders' discursive constructions of work, looking particularly at how the ambiguities of difference are articulated through work talk.[5] Secondly, we interrogate retro retailers' work biographies and how these have evolved through time and space, looking specifically at questions of flexibility, knowledge, embeddedness and risk. This focus on the employment biographies of retro retailers offers a way in to understanding the spatial and organizational evolution of retro shops and stalls and the creative quarters in which they are located. This focus also enables us to offer a more critical take on the evolution of the creative industries and on the role of small firms within it. We consider in particular how traders navigate their way through the intensifying economic pressures of the marketplace and in so doing attempt to redefine and reinscribe the boundaries between the 'mainstream' and the 'alternative'. Finally, we draw on field notes from our observational work in a range of retro spaces as a means of offering a third 'take' on the spatialities and temporalities of retro retailing.[6] This approach, we argue, has the potential to make two key theoretical interventions. First, it will extend our understandings of both creative work practices and the role of the cultural industries as tools of local economic development and place-making. Secondly, such work may offer insights into temporal reconstructions of the alternative/mainstream border and to our conceptualization of the inherently slippery and unstable spatialities of centres and margins, cores and fringes.

THE CREATIVE WORK MODEL

In the following discussion we explore the working lives of retro traders in the context of three key conceptual concerns which, together, comprise what might be termed the creative work model of self-employment. We will evaluate the extent to which such concerns lend themselves to a conceptualization of the early stages of retro trading, going on to argue that the creative work model is, however, limited in its ability to account for the more complex picture revealed by the work biographies of those who achieve commercial success as established retailers. First, we address the extent to which creative work is flexible, fluid, free and empowering; secondly, we consider recent claims that creative work is rewarding by virtue of its embeddeness in social networks which exhibit common communities of practice; and finally, we explore the extent to which retro work specifically, and cultural work more broadly, might be defined as risk-laden and precarious. Taken together, these characteristics have come to symbolize the creative work model, an idealized set of working practices centring around reflexivity, control, imagination and creative freedom which are seen by some as the key components required to meet the challenges of the twenty-first-century knowledge economy (see Burton-Jones, 1999; DTI, 1998, Leadbeater and Oakley, 1999). We conclude by raising some concerns about the ability of the creative industries to maintain their critical commercial edge and argue that the economic and cultural position of retro trading is inherently unstable as it depends on high levels of self-exploitation and on eminently plagiarizable market conditions. While the barriers to entry are low in economic terms, in cultural terms commercial success is an altogether more problematic notion which requires constant innovation and distinction.

Flexibility, fluidity, freedom, fun

In a number of accounts the cultural industries, and particularly those based around new technologies in one way or another, including multi-media and internet services, are envisaged as vibrant, imaginative and money-making. Such organizations offer the potential at least for more reflexive, individualistic social identities. They certainly open up new spaces for the creation and display of symbolic cultural capital. At the very least, it is argued, such micro enterprises enable people to search for a more rewarding job for themselves, a job over which they have a degree of control. So that as risk and uncertainty become more endemic, so people become more reflexive and more engaged in participative practice (Lyotard, 1984). This represents a 'new form of individualism whereby people fall back on their own resources to construct their own employment biographies, negotiating the hazards and opportunities in inventive ways' (Allen, 1997: 184). For du Gay, the discursive construction of work-based subjectivity and the new emphasis on creativity in work produces a new kind of worker, one who sees work as gratifying and who 'enjoys the dream of creative satisfaction in work and also the "fantasy of entrepreneurship"' (du Gay, 1996: 178). McRobbie has argued further that cultural workers adopt a creative identity that fuels their ambition and justifies the 'long hours, low pay and even the lack of success (i.e. misunderstood by critics and public alike), and enables them to tolerate, for example, wages and hours that no employer could reasonably expect'. (McRobbie, 1999: 14). Some, it is argued, find that a trade-off of autonomy against insecurity is more attractive than working for a large, impersonal organization (Leadbeater and Oakley, 1999: 15). And it is undoubtedly true that many of our interviewees explain the rationale for their involvement in creative work in terms of personal fulfilment and freedom. Work is seen by some as a 'labour of love' (Clara, Nottingham); rummaging through markets is described as a creative process (Fiona, Portobello); sourcing is described in terms of committed rummaging (Maggie, Portobello). Such jobs, it is argued, offer personal satisfaction and fulfil particular desires at particular moments in traders' lifeworlds. This desire to find a space to work independently emerged as a powerful theme throughout our interview work and is the classic explanation for self-employment, as Roger (Nottingham) explains:

L: The thing a lot of people are saying about this work is that the last thing they'd want to do is to work in a big organization with somebody telling them what to do.

R: It's horrible. Somebody looking over your shoulder all the time. Nah. Elli'd go crazy. She couldn't do it. ... If you want to meet interesting people and not have somebody telling you what to do it's great. Like Elli said when she started, she could never work for someone else. So she started it. ... There's nothing finer than being here for me. I can sit and listen to music all day. I can put any kind of music on I want, from Chet Baker ... I can put Prokoviev on, I can have some punk on. Whatever. I can sit and read. What more can you ask for?

Many retro traders argue that creativity is their key competitive tool and often reveal a social, political or personal commitment to these new forms of working. Sara (Notting Hill) is an interesting example of someone who is committed to a creative vision of self-employment and speaks about her shop as 'my dream', where 'the pleasure is me being surrounded by things that I indulge in. That's the bottom line.' She began doing the markets in the early 1980s when she was a student in need of extra cash, and quickly realized the attractions of self-employment: 'It worked really well. I enjoyed it. I loved it. It's a very addictive thing to do.' Such work is seen as pleasurable, as a means of exercising control over one's life and fulfilling one's creative potential. It is, importantly, acknowledged not to be a particularly lucrative profession, but is one where one can pursue (and fulfil) one's fantasies. Sara (Notting Hill) is again illustrative of this negotiation between money and love:

> Work is my passion. I work very hard. I'm here seven days a week, it's a non-stop thing which is not ever going to make me a fortune, so it's essentially about pleasure. And I feel very lucky. I feel blessed that I'm able to do something that I love. ... This is my dream. I'm not somebody that wants to climb rungs of ladders.

Cathy (Portobello) is revealing about the attractions of self-employment. She has a long history in the fashion industry but felt that working for chain stores such as Oasis and Warehouse was sapping her creative spirit:

> I've been in this unit about 4 months after a career of about 15 years in mainstream fashion. ... I went travelling for a while because I was so sick of working for suppliers to Warehouse and Oasis. We were just copying the catwalk all the time. It was very restrictive what you could do. It wasn't very imaginative. It was just a question of seeing other designers and copying it. I was a designer but was given such a narrow brief ... I just wasn't enjoying it any more. I wanted to get out and do my own thing. ... Having been in it for so long I just got bored and wanted a break.

She thus gave up her formal designing job, indulged her pleasure for travel, picking up design inspiration on the way, and began doing the markets on Portobello Road. She moved into her shop because her stall was doing so well and 'sold out in three weeks'. She is very clear about the attractions of self-employment, emphasizing the trade-off between money and pleasure:

> I've never been happier. To be honest, when I was working before I was earning twice as much working half the time, but I wouldn't swap it for the world. ... I don't want an empire or anything like that. I want to keep it small.

And again Roger (Nottingham) summed up the pleasure of working for oneself in spite of potentially poor financial rewards:

> I couldn't work. I can't see why anyone would want to go to work, I'd go crazy. ... But if you want to be a millionaire forget it. ... It's certainly better than working for a living. Money's not everything.

What this points to on first reading, then, is the emergence of a new kind of subject for whom work is understood in terms of individual creativity (du Gay, 1996). This echoes McRobbie's accounts of the new media and culture industries which depend on a fusion of entrepreneurial values with a belief in the creative self (McRobbie, 1998: 83). Certainly, listening to their own accounts of their work biographies, the creative entrepreneurs whom we interviewed appear committed to holding out for a life which meets their personal aspirations. Their talk emphasizes the desire to make work something more than a mindless drudge and repetitive chore, while acknowledging that their business is unpredictable and prone to transience.

However, while self-fulfilment, freedom and creativity are undeniably important factors in explaining the attraction of such work, we would argue that those traders who become sufficiently financially successful to set up more permanent shop outlets as retro traders begin to show a more ambivalent relation to work and articulate both how the trade is becoming more competitive and how their relationship to their work changes, as we go on to discuss below.

Communities of practice, embeddedness and gendered social networks

The second set of reasons which are cited in the literature to explain the attraction of creative work relate to the ways in which such work is embedded within social networks which strengthen and support what might otherwise be solitary and isolated working practices. And again, on first reading, the importance of informal social networks were seen by many of our traders as an important factor in accounting for job satisfaction. Work becomes a fun, sociable activity where you are surrounded by, and interact with, like-minded friends and colleagues. As one experienced trader argued:

> K: You know lots of people and everyone knows everyone else and what they're doing?
>
> P: Oh yeah, it is in Bristol, cos we used to do the Canon's Marsh boot sale, we both did, and we didn't know anybody, and we know loads of people now, cos they're all into the 50s, 60s, 70s, or they're interior designers or shop owners, yeah, Canon's Marsh used to be our social life. We just sat around chatting in the afternoon.

Similarly, Elsa (Nottingham), a recent entrant into the trading scene, talks of her workworld as one of an elite, 'knowing' group (including other designers, shop owners and those involved in 'approprlate' club nights) that outsiders (people in the street) wouldn't understand. Sara (Notting Hill) also talks about the sociality of Portobello, where people

> come by and say hello. That kind of thing. There was a sharing of ideas. The people who came to buy tuned into your tastes and shaped your tastes. You kind of feed off each other, that kind of thing. That's quite interesting. Quite inspiring. ... I have a lot of friends here and we go out socially. I don't really go out of this area much at all. It's good, it's self-contained, it's got a little nucleus.

This mix of knowingness and social connections as a way of creating a scene is typically fostered on friendship networks and embedded in circuits of social and business networks. It relies on certain communities of practice and knowledge – about where to go, where to meet people, what's in and what's not. Such economies of proximity, where work is carried out in bars, clubs and restaurants as much as in the formal workplace, go a long way towards explaining the attraction of self-generated self-employment. One of the ways that retro traders explain their work is thus by aestheticizing both the place of work and the time of work. For many, the shop becomes a creative space and work-time becomes expansive, the long hours slogging into the night to source, produce or alter goods, or the nights out spent chatting about 'work' becoming glamorized, fun. And this brings about a curious reversal between the world of 'work' and the world of 'home'. Hochschild explores this emergent world where work becomes 'home' and where work ties replace kin ties, friends replace family. Although Hochschild's work relates to large American corporations, her exploration of the transposed world of home–work bears remarkable similarity to the accounts told by our retro traders, and gives us some purchase on new rearticulations between work and identity. For Hochschild, work – not just the substance of it, but the buzzy surface feeling of working life – is for many a source of pleasure. She talks about the overtime hounds who crave extra shifts, the workers who love the job because they can work at Christmas. She talks about a world where work becomes like a surrogate home, a reversed world, a world where people give everything to work. Like our interviewees, Hochschild's study explores the people who feel that the true centre of their social worlds is work, not the neighbourhood. Where life at work is, frankly, more fun, and where self-satisfaction, well-being, high spirits and work are inextricably linked (Hochschild, 1997: 41). However, while this is the ideal form of work among our retro retailers, it is often not achieved in practice and is seldom the eventual reality of their working biographies. So far, then, the existing model of creative work seems to describe retro trading practices well. However, as we have argued, such a picture appears limitingly definable, fixed and static, and fails to take into account the complex ways in which such work practices evolve through space and time (see below).

Precariousness and risk

A final component within emergent literature on creative work, and one which again raises questions about the usefulness and accuracy of the model, is the suggestion that the creative industries are volatile, transitory and inherently risky. A number of commentators have argued that work is becoming more precarious and less predictable (Allen, 1997; Allen and Henry, 1997; Beck, 1992; Beck et al., 1994; du Gay, 1996). Beck has discussed the emergence of a risk-laden system of 'flexible, pluralized, decentralized underemployment' (Beck, 1992: 143) which throws up new forms of uncertainty and unpredictability. He argues that 'abnormal work', with its unpredictable and erratic rhythms, is now becoming the norm for increasing numbers of people. Under such circumstances, identities become both fluid and fraught, dislocated and ungrounded in former certainties about 'jobs for life' and full employment. In the specific case of the creative and knowledge industries, it has been argued that we are living on the 'thin air' of knowledge production, application, exploitation and dissemination (Leadbeater, 1999); in a world where we feel more uncertain, stressed and insecure. Other creative sectors are, in contrast, seen as volatile, unpredictable and unstable, falling into the classic small-firm syndrome of under-capitalization and fragility (McRobbie, 1989).

While in much of the literature risk is seen as endemic within micro- and small independents who face new and uncertain markets, unforeseen competition and unknown costs, our study of retro retailers' biographies suggests a rather different take on risk. We would argue that in many cases retro trading is a comparatively risk-free option. The barriers to entry are minimal, as there are few or no training implications and little need for extensive property investments. Commitments in terms of opening hours are limited and can be varied, and sunk costs are negligible (many enter into temporary lease arrangements in otherwise vacant units). In the absence of any alternative 'career' trajectory, the only thing at stake, it seems, is the potential to break one's dreams. This is, in many respects, a low-risk strategy, an alternative form of (often temporary, typically part-time) employment. To illustrate these arguments we again refer to retro traders' work biographies. Many reveal highly flexible opening hours and work seemingly to suit themselves,[7] as Pete explains '[Saturdays] are the only time I'm officially open, but I have been open during the week whenever I've felt like it, for an hour or two, you know'. Cathy (Portobello) argues that 'I love it here. It's great. And also you don't have to open every day of the week. People would like to only open on Fridays and Saturdays down here.' For many, trading evolved out of their own collections; it was rarely part of a career trajectory or business plan and was more often than not the result of luck, chance or friendship networks. Joan (Affleck's Palace), for example, reveals how she 'did the odd flea market and I just got interested in it'. Vinnie (Bristol) similarly tells how it all started from jumble sales ... when the room was so full of stuff collected through the year going to jumble sales ... we decided to open a little stall'. The important point about such biographies is what they say to us about risk within this particular sector of the creative industries. Specifically, we would argue that this kind of work does not strike us as risky at all – as Max argues, you get a key and you open up. So what this sector offers us is a new take on questions of employment flexibility and risk. It also suggests that we need to look not to questions of precariousness and risk, but to more conventional economic geography questions of rent, rates and commercial investments to understand the emergence (and demise) of creative quarters.

RECOMMODIFICATION AND THE TEMPORALITIES OF RETRO

Analysing traders' work biographies reveals a markedly shifting temporality as the retro trade has become more competitive in the past 10–15 years. Maintaining a sense of distinctiveness and difference has become increasingly difficult as knowledge becomes more widespread, as the market becomes more crowded, as new forms of commodification challenge consumption preferences and as property (re)valorization becomes a prohibitive force for small independent traders. These trends in turn have marked organizational and spatial implications.

Property revalorization

First, there is the problem of property revalorization as formerly marginal spaces undergo economic regeneration and become more desirable and popular. Part of the initial attraction of de-industrializing urban spaces for many retro traders and – subsequently – retailers was their affordable down-at-heelness. Cathy (Portobello) explains the ease of entry into self-employment as a fashion designer which the designer units at Portobello Green offer:

L: Is it fairly low start-up costs here?

C: It is, yeah, it's really great. I wouldn't have had the confidence to do it otherwise. Here it's only a notch or two up from doing the markets. You can afford to make a few mistakes. Take a few risks. It's excellent, it's really good. I definitely wouldn't have had the nerve to go straight into a big shop myself with the rates they charge there, and you're often tied in for a long time. But here it's ... not much of a commitment ... and you don't have to open every day of the week.

Bureaucratic arrangements and economic tie-in clauses rarely figure in emergent creative quarters, and the ease of market entry and exit typify such spaces. In the case of retro trading in a market setting, the barriers to entry are minimal and risk really isn't an issue. In the case of Portobello Market, for example, the market structure of low rent and direct contact with customers, designers and cultural entrepreneurs enables costs to be kept low and personal investment and risk to be minimal. Similarly, the low rents and ease of exit at Afflecks Palace in Manchester make it a good seed-bed for new small firms, a launch-pad as Leanne the manager there argues, a starting-off point for local design talent. Leanne and her partner John further enable entry into this marketplace by adopting a nurturing attitude towards new stallholders, helping with the design of creative interiors within Afflecks Palace and guiding new entrants through the more complex bureaucracy of business start-ups, including tax advice and business management. Max, a former Afflecks Palace trader and now a successful and high profile retailer, endorsed this relatively risk-free view of the space:

It's quite a good way of starting a business because it's one old warehouse and multi tenanted, and if they like your face and what you do and they think you can add to the mix of the place, then you can have a unit, and there's a single weekly payment which covers rent, rates, electricity and everything which makes life quite easy and means you don't have to take a lease and get involved with solicitors and all this kind of thing. ... [It's] quite a good system, certainly for starting out.

However, and as we go on to discuss below, the popularization of creative quarters ultimately leads to an increased commercial interest in such areas and to spiralling property prices which threaten to push out the original creative pioneers. Max explains this process in the specific case of Manchester's Northern Quarter:

What happens in regeneration areas in towns and cities ... particularly in this area of Manchester ... they tend to be occupied by creative industries. You get a gap basically. The Arndale Centre left this area vacated, so you've got a lot of experimental businesses and artists [moving in] until eventually there's enough critical mass for other people to start getting interested. And what happened there was that Afflecks Palace was one of the earliest businesses actually, and then people like me [who] start there, move out of it and open up other places [which] promote[s] the area. You start attracting more people which keeps you going and that starts expanding and then eventually people's rents get assessed

> and people start going out of business. It's a fantastic spiral but it's what happens. You either don't do anything and stay at the bottom of the dereliction pile or you do something and just adapt and move on. ... [It's] the landlords who are most greedy and the ones who cause the problem.

Leanne, the manager at Afflecks Palace in Manchester, also discusses the problems of urban regeneration. She describes how the Northern Quarter area was a largely derelict zone of seedy sex shops and pawnbrokers following the relocation of the Arndale Centre in Manchester,[8] and tells how she and her partner battled with an initially hostile local authority who wanted to close Afflecks down. But as formerly derelict or down-at-heel areas become more popular, so rents and rates are pushed up in the classic urban story of gentrification, as Max (Manchester) reveals:

> The lease thing can be a bloody nightmare, especially when you get your rent reviews and the rents go up and life has changed and moved on. That's happened here you see [Café Pop]. We opened here about 5 or 6 years ago and [now] they want to treble [the rent] which of course puts everybody out of a job. But it's what happens you see in regeneration areas.

Here, then, we have to look towards far more conventional urban geography explanations about property rental values to understand the origins (and demise) of creative retail quarters such as Camden and Portobello in London, the Lace Market in Nottingham and Manchester's Northern Quarter.

Sourcing scarcity

Secondly, as the market becomes more crowded, sourcing original goods becomes increasingly difficult. The problem is that genuine articles are becoming more and more scarce, and more and more expensive. As Sara (Notting Hill) argues:

> L: Has sourcing become more difficult recently?
>
> S: Much more so. Things are becoming incredibly scarce. It's getting harder all the time because of a sheer lack of things. We just don't see much vintage around these days. As the years go on the condition gets worse and the challenge gets harder.

Max (Manchester) also acknowledges the sheer hard work involved in sourcing retro commodities:

> I do very little on the buying side now, I more or less stopped about a year ago. ... I've had the most bizarre channels. There is no good place

> to buy it. Car boot sales are not a good place to buy it. You traipse round seven car boot sales in the pouring rain for 6 hours and maybe you'll buy three or four things. It's an endless filtering process so you get very little. Antique shops tend to know nothing about anything whatsoever. Junk shops and charity shops, you get very little in the way of quality. ... It takes forever basically.

The difficulties in sourcing original items and the inability to predict supply or to match supply with demand are the reasons Max (Manchester) cites for his current disillusionment with retro retailing:

> In a nutshell, the reason it doesn't work as a business is that it's terribly time consuming to get the stuff; it doesn't really work as a retail business anyway because people come in and want specific things and you haven't got it and you can't tell them when you're going to get it.

These difficulties, then, lie at the heart of the current dilemma facing retro retailing, and give rise to a sense of nostalgia about when the trade was easier. First, certain retro retailers reveal nostalgia for an earlier, and possibly idealized, time when retro stock was apparently plentiful and when the market was uncrowded. As Sam (Nottingham) commented:

> I'd go and buy the most incredible things ... it's nice to buy '60s things because its one of the last things around you can buy relatively cheaply, easily ... it's all different, abstract, good quality ... at the time there were very few other people doing that. ... That was in 1989; there was another shop in Manchester and maybe two or three in London doing post-war stuff.

Similarly, other retailers talked nostalgically of a time when retro trading wasn't about big business but was about being part of an innovative, knowledgeable scene that was setting the trends for the alternative market: 'my business was about doing things different' (Max, Manchester). Finally, there was a lament for when youth (particularly student) culture was about looking alternative: 'schoolkids just want to look like everyone else today. ... I despair' (Roger, Nottingham); 'that generation now at University ... wanting ... Calvin Klein' (Tom, Bristol).

Situated knowledge, tacit skill: distinction, taste and the problems of staying ahead

Thirdly, and partly in order to counter the difficulties of sourcing good-quality items discussed above, retro traders and retailers depend on a repertoire of knowledges about the retro trade. Analysis of traders' biographies in turn reveals both the importance of situated knowledge in ensuring commercial distinctiveness and the difficulties in renewing and sustaining such knowledges over time. In certain senses a reliance on the competitive tools of situated knowledge and tacit skill, which are innate, instinctive and not easily learnt, is one of the ways in which our interviewees made sense of their work biographies. A number of our interviewees have a long history in fashion and many have moved across into retro trading and retailing from other creative sectors such as music or art. Sara (Notting Hill) trained as an art historian and was 'always interested in the visual, aesthetic, creative side of things'. Similarly, Max (Manchester) has a long-standing interest and involvement in design, revealing how he:

> Started originally 11 years ago in Afflecks Palace down the street ... and I had a small shop; actually I was a design student originally and I became interested in the social history of the twentieth century related to products, quite a high brow interest. I started a few years ago, collecting icons of their day, design icons. It was an obscure thing to be doing at the time and it expanded and became a bit of an obsession. And I couldn't really think how I would be gainfully employed either, being a student, so I decided perhaps I could make a living out of buying and selling design items, second-hand goods basically.

Again, Sam (Nottingham) began in a band and supported himself by doing markets. He was always, he argues, 'interested in design and clothes and things, that kind of thing'. Pete (Manchester) had been to Art College and began a stall in Afflecks Palace, while Elsa (Nottingham) trained as a designer and worked in the clothing business for some years before setting up her retro business, and Matt (Portobello) is a trained architect with a sideline interest in retro and antiques.

Others have developed situated knowledges not through formal training in the cultural industries, but rather through long-standing family or personal histories of buying and collecting second-hand commodities. Clara has a long family history in antiques and remembers how she used to go to auctions and fairs with her mother, looking for Victorian lace and linen. Others, such as Elli, have worn second-hand clothes for 20 years and have developed an eye for sourcing commodities which will sell. Elli's partner, Roger, tells us 'don't ask me how she knows, she just does. ... She's got an eye for it. It's just a knack. You've either got it or you haven't. You can't teach somebody'. Similarly, Pete (Manchester) reveals how the knowledge to spot the next trend is an unteachable skill:

> Once something becomes valuable, you don't see it ... that's the fascinating thing, seeing what's going to be the next big seller ... it's not predictable. Something can be laughed at and then two years later everyone wants it.

And Tom (Bristol) argues that 'it was interesting to see how some people understood it [retro] straight away and other people, however hard you tried, never actually grasped the concept'.

This, then, is perceived to be an instinctual talent, a gut-feeling, a sixth sense about what future likely trends will be. It is inherent knowledge but, crucially, knowledge that can be shared and appreciated between different retro retailers. It is a set of knowledges that depend on being immersed in a creative scene where emergent trends can be both identified and shaped. As Sara (Notting Hill) explains: 'I get a lot of inspiration from the clothes coming through the shop. ... You kind of feed off each other, that kind of thing.' In part, then, such situated knowledge is deeply geographical, grounded in particular places at particular times and emerging through the activities of groups of people who become identified as pioneering, innovative, in touch with fast-paced shifts of fashion and style. Such geographically situated knowledge goes some way towards explaining the success of second-hand and retro trading such as Portobello Market where designer-traders such as Toni talk about how easy it is to spot trends just as they are about to take off. Without this, she argues, designing would be impossible. Such insider knowledges, the metaphorical nudges and winks between accomplices and the silent, hidden references and codes, create a kind of secret language of distinction (Bourdieu, 1993) which confirms a complicity between the taste-makers and excludes the layperson who is always somehow bound to miss the point. Sam (Nottingham) is interesting here when he discusses how he sells industrial designer ware which many people don't know about:

> Things designed by Olivetti or Braun, that kind of thing. A few people do know about it but your average Joe doesn't ... to most people it would be like an old typewriter, old calculator, old phone, whereas certain people know what it once was.

Bourdieu explains how the new taste-makers expect images and representations designed for their consumption to require the play of specific codes and competences. Here we see the importance of distinction through second-hand consumption, where innate knowledge and cultural capital enable entry into a secret style world where there's a wealth of knowledge, but it is knowledge which can't be learnt or taught. And such situated knowledges are one of the means by which people carve out a niche for themselves in an industry in which making one's name so often relies on marking oneself out as different from the mainstream or dominant form of organization.

But the difficulty with relying on tacit knowledge as a competitive strength is that aesthetic and stylistic shifts mean that knowledge is constantly under review, is constantly in flux and is thus elusive and difficult to stabilise. And so the business of retro retailing in fraught with difficulty as it depends on eminently plagiarizable notions of quality, distinction and taste. To stay ahead of the game in an ever more competitive style arena requires very specific forms of knowledge and skill which depend on the definition of difference, a project which is becoming increasingly difficult – and within which one's position as 'alternative' retailer becomes increasingly ambiguous – due to the recent commodification of retro, as we go on to discuss now.

Retro recommodification: reappropriation, assimilation and the erosion of 'the alternative'

As the mainstream adopts retro style through the sale of new fashions based on copies of originals, retro traders and retailers face damaging competition from more conventional retail spaces. As we have already argued, the present problematic negotiation between sourcing and selling original goods while staying in business and fending off mainstream copycats prompts a high degree of nostalgic talk about when these tensions were perceived as being not so apparent. Most of the traders we interviewed began trading in the early 1980s. One significant theme in our interviewees' descriptions of the time is that in the 1980s traders and shoppers were both part of an alternative, elitist scene, in which commodities were traded and consumed according to the then appropriate values of vintage, authenticity, difference and individualism, and which took place in appropriate alternative consumption spaces. Camden, Portobello Road, Afflecks Palace and Park Street in Bristol all became significant alternative shopping spaces for retro in the early 1980s. In the 1990s, we saw more of an assimilation of retro into mainstream culture as the 'alternative' tastes of this elite and knowing group trickle down to wider groups. Possibly this is part of an inevitable cycle whereby what was once an alternative, anti-good taste, anti-high fashion statement becomes assimilated into mainstream culture. So as the alternative traders of the 1980s become more established, and more mainstream, the boundaries between retro and the high street become blurred. The final stage of the cycle is when this trend becomes so widespread that mainstream retailers cash in on it. Thus retro has now become a dominant fashion story on both the catwalks and the high street, and the cycles of reappropriation are speeding up and spreading out, such that both the 1970s and the 1980s became dominant style motifs during the late 1990s. We would argue that for retailers of retro, this simulation of the past, this reproduction of the authentic by the high street lies at the heart of their current dilemma as they see their businesses being ruined by 'inappropriate expansion'. The organic scene should be allowed to develop at its own pace argued one interviewee, not be forced by media attention and popularity, 'the whole thing has gone a bit topsy turvy ... people are [cynical], and that's what spoils it' (Sam, Nottingham). Repro-retro is, argues Sam,

> not the thing. It's just doing kind of reinventions of it. The design quality, the quality of the products is poor and doesn't last; it's got quite a short lifespan, hence it's cheap. And I think the quality of the design is poor and usually I can see what makes a watered down version and the lines aren't quite right and the colours aren't quite right. ... They're all imitations ... so I'm a bit snobby about it and don't rate that kind of stuff at all actually.

Original traders thus become disenchanted and disillusioned about this trend, given that this is precisely what they set out to avoid. Max (Manchester) goes on to argue that,

> Once everything becomes mainstream you've already lost it. Manufacturers are manufacturing bad versions of originals which completely devalues the original item ... and that's very much what's happened now really; everything's become a mish-mash of fifties, sixties and seventies in a nasty kind of retro.

Echoing Max, another interviewee discusses at length how such expansions of the 'alternative' marketplace, in this case Camden, has 'diluted' the whole experience, making it touristy – and impossible to get 'real' bargains (Sam, Nottingham). The market 'has changed' argues Sam, 'it gradually just got bigger and bigger. It's difficult to do and actually make it worth doing, especially nowadays.' In contrast to the somewhat idealized accounts of the traders 'feeding off each other' discussed above, this increased media and popular interest in creative quarters such as Notting Hill, Camden, Nottingham's Lace Market and Manchester's Northern Quarter is seen by retailers as forcing them to re-evaluate and reassess their trading strategies. Max

(Manchester) argues, for example, that you either have to either ignore market pressures and stay true to your ideals, or move with the scene and adapt accordingly,

> You either go, 'Fuck it', [if you are doing something unfashionable] and carry on regardless, which is what I've always done, or you endlessly adapt and change if you can do. I mean basically if you can just add more and then keep moving with the time as it were, which is just hot on your tail, and keep adding more things. It all comes back [into fashion] at various points. ... There's a revival of everything in the end, I kind of wish there wasn't now, I think it might be time to move forward.

Others acknowledge their early naivety and accept that their early impressions of the retro world have, in many instances, been tempered by less aesthetic and more commercial concerns. Sam's biography (Nottingham), for example, reveals this shift through time and space as experiential learning takes over from early enthusiasm and idealism. Sam began trading at Chalk Farm Market in London in the late 1980s, sourcing from car boot sales and amassing large amounts of stock in his house. His work narrative describes the market then as very underdeveloped and himself as one of only a few relatively successful traders. He then moved on to the provinces, and set up formal shops in creative quarters in Manchester and Nottingham in the early 1990s. His first shop was in the edge-of-town student quarter of Forest Road in Nottingham, a 'relatively cheap location'. He then opened a shop in Nottingham's Lace Market, and again underscores the importance of economic rent in defining alternative quarters: 'we don't have a lot of choice when it comes to paying rent. We need a lot of space and city centre rents are astronomical.' He currently owns and manages three shops in Nottingham's Lace Market, the most recent of which is a large, conspicuous and highly stylized interiors shop on a major thoroughfare adjacent to the newly constructed and high profile Nottingham Arena. It would be difficult to imagine a less alternative location and store space, and Max is well aware of how market changes have forced such spatial and organizational shifts towards more formalized and conventional trading patterns. For retailers, then, mainstream assimilation is seen as inevitable, even if it is initially undesirable.

ALTERNATIVE SPACES?

We turn now to arguments about the spatialities of retro retailing and its relation to 'the alternative'. The sense of chain store encroachment into formerly unique areas discussed above was also reinforced throughout our observation work. Nottingham's Lace Market now has the ubiquitous chains Café Rouge, Lloyds and Café Metro, cities such as Brighton are becoming disappointingly mainstream and Covent Garden is more of a tourist attraction than a market-space:

> Brighton was a bit of a disappointment. [It's] become a victim of its own success: as more trendy young Londoners move here, the place becomes more of a London suburb-by-the-sea. Certainly it didn't have many decent second-hand clothes shops [just] expensive 'designer seconds' shops. The whole thing has become more 'professional' so now we have either designer or vintage, a general professionalizing process that happens as second-hand stalls gain a reputation. (KB notes, 11 September 1998, Brighton)

> [I] got the distinct impression that this shop was now on the fashion tourist trail of London. I couldn't help but be pretty disappointed overall. I couldn't help but feel that I was on some sort of prescribed alternative shopping tour, and that these fashion-guide stickers [displayed on windows of shops featured in Guide] codified this. (NG notes, 17 August 1998, Covent Garden)

> The place was mad with pre-Carnival hysteria although most of the activity seemed to come from European tourists clutching London A–Z's and searching for antiques stalls (all of which were of course closed until Friday). And while there was a real buzz to the Notting Hill area, much of it was coming from the new and outrageously expensive designer boutiques rather than from second-hand or vintage shop. (LC notes, 24 August 2000, Portobello)

Sara (Notting Hill) also talks about the pressures towards mainstream conformity and explains how Portobello is becoming so much more commercialized following the launch of the film *Notting Hill* (1999) Starring Julia Roberts and Hugh Grant:

> L: Have you noticed the area changing from when you first started doing the markets?
>
> S: The market's changed a lot, and I don't think particularly for the better unfortunately. It's more commercialized and touristy which is a shame because it was somewhere that was always renowned for having character and new ideas.
>
> L: How has it changed? Is it since the film?
>
> S: Yes, very much so. I curse the day that film was made. It's a real shame. It's brought the wrong sort of people, the wrong sort of money injected into the area and its ousting out a lot of people. Rates are going up so small businesses can't survive. All the chains are coming in. It's just defying the whole point. It's hideous. Absolutely hideous, with Starbucks and Café Rouge.

She does, however, feel that the long fashion history which Portobello has and its reputation for design inspiration will remain, and that media attention will soon move on to the next up-and-coming place:

L: Do you think that Portobello can maintain its sense of vibrancy?
S: I don't think it will ever go. Portobello's been here too long. Its roots are too deep, but its being chipped away at, definitely, which is sad.

So for those like Sara and Cathy in Notting Hill and Max in Nottingham's Lace Market, whose work is immersed in an area with a long-standing fashion tradition, there is a sense of optimism and they have successfully made the transition from irregular market stall-selling to retailing from more permanent, more commercialized shops in more conventional locations. Others, who are in less fortunate locations or who were perhaps less prepared to compromise, have suffered the vagaries of volatile competition and have adapted to ongoing pressures towards mainstream conformity in a range of different ways. Some, such as Clara (Nottingham), have moved back historically and retreated into more vintage and antique commodities. She argues that she could not survive by selling 1970s stuff alone, and that diversification of her product range is absolutely vital. She has, for example, just bought thousands of pairs of 1940s and 1950s stockings which are proving to be enormous sellers. Others, such as Si in Covent Garden, attempt to head off intensifying competition and to play conventional retailers at their own game by introducing new ranges into his formerly exclusive retro store. Mass participation and the mainstreaming of retro serves to push high investor, creative-led traders away into something more knowing and discerning, but their creative edge is constantly under attack. A final group of traders has moved on (into new jobs, new sectors, new lives) and closed down. As Max (Manchester) reveals:

I'm getting out of it now, I don't want to do it anymore. I'm selling my own collection, hopefully to a gallery or museum so it becomes a public collection. I've done so much I think I've lost the plot now!

Our field observations also revealed the large amounts of work which go on behind the scenes: Cathy (Portobello) spends her days at the back of the shop sewing/embellishing/ironing stock unless she has a customer in; Roger, Elli, Clara and Elsa (Nottingham) all sew, alter, clean and iron commodities prior to their sale; Clara (Nottingham) scours the obit columns and trails off on (often unfruitful) sourcing journeys to people's houses; while Fiona (Portobello) spends her week washing and cleaning stock, ironing and pricing everything up, getting up at 5am to load the van, and is currently designing and redecorating the interior of a new shop. And so while the typically articulated 'ideal' trading situation involves a shared, organically evolving set of values and ideals, which spin out of mutual connections and contacts and are shored up by an identification of retro retailing as an alternative and more fulfilling way of working, this is in many senses an overly romanticized vision of second-hand trading which belies the hard work, boredom and commercial crises which typify creative work and which begin to erode the ragged imaginary boundary between conventional and alternative work. Ultimately, it seems, this kind of creative work can become just as tiring and dull as jobs in more conventional sectors of the economy, and what begins as a route into a self-fulfilling work world ultimately becomes tiresome and too much like hard work, as Max (Manchester) explains:

> I've lost interest and it's all a bit shoddy now. I'm rather tired of it all. It's all really old hat to me. ... I still go to antique fairs and get really excited about one or two things but it's just not there as much as it was, and I certainly don't want to make a living out of it anymore, it's too difficult.

Again, our observation work was suggestive of long hours spent sitting at a till in an empty shop waiting for non-existent customers:

> Mainly this afternoon I'm struck by how bloody DULL it must be to run a stall. In fact, given the image of the place and the general assumption about such jobs – working for yourself, doing something creative and alternative, not doing a 9–5 job, being able to dress/do what you like. The cold reality is it is very, very boring most of the time. I hear one woman when asked if she's enjoyed her time off: 'I was bored! I didn't think I could be more bored than sitting here all day but I was!' (KB notes, 21–22 January 1999, Afflecks Palace)

And again,

> Once again (as in Afflecks) [I] am forcibly struck by the contrast of how people speak about working in this environment (creative, exciting, doing your own thing) and the utter tedium of the reality of sitting at your unit, bitching with the person next door, also slumped over the till, waiting for the students to get up and come in. (KB notes, 19 February 1999, The Forum, Sheffield)

Different traders and retailers are, we argue, producing these spaces in very different ways and their understandings of, and investments in, particular decades are used to construct a range of spaces of second-hand exchange which all, in various ways, sit uncomfortably alongside an imagined mainstream 'other'. Many shops, for example, become almost personal galleries or exhibition spaces, reflecting their owners' aesthetic ideals, displaying commodities which they have either designed, made or selected, and constructed to attract a particular clientele. Sara (Notting Hill) is a clear example of this tendency:

> It's about beautiful things. Eclectic beautiful things that are there for style, cut, quality of fabric. It's really appealing to anybody who wants to tune into beautiful things and is looking for something a bit different. Like me. That's what I seek when I go shopping. So it's providing the aesthetic beauty, presented well.

Similarly, Elsa (Nottingham) would prefer to sell to people like her – the knowing who appreciate her clothes and understand what she's about. This is about creating a buyer–seller relationship which goes beyond the dynamics of exchange and which contains facets of the gift relation in that commodities are carefully selected and displayed to appeal to particular kinds of consumer. Elli (at Baklash in Nottingham) presents her shop as a creative space quite literally by using its interior walls as display space for young local artists.

Extending this, these and other such stores endeavour to mark themselves out as different from conventional retailing through the employment of distinctive spatial tactics. In the case of a market setting such as Camden, Covent Garden and Portobello, such strategies are comparatively easy given the nature of the sites, which are temporary and outdoors, and thus by definition different from conventional retail spaces. Our observation work describes the busy and boundary-less area of Camden:

> Lots of skinny trendies wandering up and down and 'Being Seen'. The shops leading up to the market itself are set out like market stalls – they are open-fronted, with loud music playing, stuff hung up at the front and displayed on the pavement so it's not always easy to say where one shop ends and next door begins, some shops become a different shop in the basement. So shops' boundaries are creatively blurred here, and also the boundaries between the market and the shops ... shops and stalls are places where people socialize, and nearby cafés. The whole place seems full of people just being seen, customers socializing with shop staff, shop staff meeting in cafés. (KB notes, 14–15 August 1998, Camden)

Other require a certain geographical knowledge on the part of consumers – simply locating certain retro shops in off-the-beaten-track places can be difficult, again as our observation work again revealed:

> It's [down] a sort of alleyway. You'd miss it if you didn't know it was there. It's closed on Mondays, unless there's a Carwash[9] that night, when it stays open. (LC notes, 13 August 1998, Nottingham)

And other retail spaces employ spatial tactics of disorder and confusion in order to differentiate themselves from formal, first-cycle spaces. This, coupled with the adoption of stall-like units which emulate a market setting, characterize Afflecks Palace in Manchester:

> Afflecks Palace is both cavernous and structurally complex. The interior seems to be creatively illusionist, and the boundaries, ·borders, entrances and exits are all very ill-defined. The stalls themselves blend and blur into one another and thoroughfares are rarely evident. This is a profoundly social place. There was no sense of surveillance, either by security people, cameras or even unit owners/workers (in fact it was often difficult to tell customers from workers). (LC notes, 31 July 1998, Afflecks Palace)
>
> The sense of disorder (chaotic displays, indeterminate boundaries between units) hit me. ... This is a place where you could quite literally lose yourself. ... I never quite knew which floor I was on, how to find the stairs, which unit I was in. ... This all seems to be part and parcel of the Afflecks Palace 'alternative' shopping experience where the rules of the game seem to be to break the rules of conventional consumption in terms of spatial layout, range of goods, price, rummagability. ... Here we see a space for fun, for irony, for kitsch, for lingering and loitering and seeking and sorting.[10] (LC notes, 31 July 1998)

This is very unlike conventional first-cycle retail stores such as Marks & Spencer and BHS[11] (and intentionally so, we would argue), which have highlighted walkways to steer the customer through the store and maximize exposure to the goods, and where the entrances, exits and pay-points are all illuminated. Clearly, then, the aim at Afflecks is to engineer a shopping space that breaks the rules of conventional spaces in terms of layout, fixtures, fittings, boundaries and product mix. Afflecks Palace seems to be consciously trying to undermine all of these consumption conventions. It is place to stop, to look and to (quite literally) lose oneself. In addition, there is a high degree of turnover in the units and one of the continual tasks is to replace stallholders. This is a highly regulated practice with stringent selection criteria, including an interview with potential traders, which are designed to screen out the types of stall Leanne doesn't think are appropriate here. The organization of space is thus stage managed and orchestrated in order to achieve an appropriate clientele and product mix. Leanne uses notions of exclusivity and practices of inclusion and exclusion to represent, and hence constitute, this space in particular ways, as a space of distinction and differentiation from the mainstream. But, paradoxically, our reading of Afflecks Palace was that it is remarkably homogeneous and uniform, and fits around a particular nexus between fashion and music which characterizes the Manchester scene. It seems to be firmly grounded in street and club wear and those involved in Afflecks are both part of this scene and also are instrumental in (re)producing it. It has, moreover, emerged as a tourist site in recent years:

> Afflecks Palace is celebrated, fêted as a tourist attraction; emblematic of trendy Manchester – and indeed when we were in Café Pop later there were a couple with a tourist map out on the table, obviously 'doing the sites' in a way very reminiscent of the marketing of Beatledom in Liverpool. It struck me that there was a LOT of this going on – people come to AP to consume it as a spectacle, just as much as they come to buy. (NG notes, 30 November 1998)

So again we see how the production of Afflecks and its representation as alternative is beset with contradictions. On the one hand it markets itself as alternative, yet it also parades itself as a model of alternative shopping and a tourist attraction. So it is, in a sense, being constructed as prescriptive, which is precisely what the alternative would normally constitute itself against. This, it seems to us, is a deeply problematic and ultimately unsustainable notion of the alternative.

Other retro retail stores have developed as playful, temporary, fun and excessive incursions, stimulated, in large part, by developments in the music and club scenes.[12] These retro spaces are constructed much more closely in accordance with the dictates of the market and are much more formalized, much easier to read and tend to be dominated by items for which demand is high and turnover rapid (1970s shirts, flares, flight bags, and so on). The Forum in Sheffield is a good example of a space which has tried to emulate a market-type setting but which has ultimately failed to offer much that is different either spatially or in terms of product mix and display:

> All in all it's rather predictable. Nothing really outrageous or different. It really does seem to me to offer a pre-packaged, pre-selected range of student-oriented gear and not much else. All in all, a little bit of a disappointment. (LC notes, 18 August 1998, The Forum, Sheffield)

In many ways such spaces are scarcely different from first-cycle, formal retail spaces and the practices, constructs and ideals of formal retailing are thus the primary producers of these spaces: profitability matters and they are a long way away from the artistic design-led imperatives of the others. Most significantly of all, we would argue, is that the balance between such spaces has shifted yet further with the intensification of the commodification of retro.[13]

These varied spatialities are, it seems to us, a profoundly ambiguous and ultimately unsustainable notion of the alternative: while such retail spaces are constructed as alternatives to the high street, their practices, commodities and social relations of exchange and consumption seem remarkably mainstream and straight. Alternative retail spaces are bounded, contained and construct themselves as alternative, but are refracted through and against the world of the high street and seemingly cannot escape it. And so as property developers and media journalists move into an area, rental values increase and creative kudos is lost.[14] And as the mainstream encroaches into formerly alternative spaces, so formerly distinctive retail arenas lose their aesthetic and creative edge. Mainstream assimilation is, as we have argued, undesirable but inevitable. And there will thus always be limits to the ability of alternative retailing to be radical and transgressive.

CONCLUSIONS

There are three broad sets of conclusions which we want now to draw out from the above discussion. The first relates to retro trading and to a set of arguments about creative work, its aesthetic and economic attributes, and its radical potential. The second relates to a critique of existing literature on creative work which is static and bounded in that it fails to take into account the shifting temporalities and spatialities of creative work. The final set of conclusions relates to retro retailing and to questions of assimilation, commodification and the problematics of boundary construction.

First, then, we argue that new working practices within retro retailing are, in certain senses, challenging many of the conventional wisdoms about self-employed work as fragmented, risky and precarious. What our research is suggesting is the possibility of pursuing creative, knowledge-led and satisfying work strategies, whose potential transience and limited scope for commercial gain are mediated by notions of sociality, trust and friendship alliances. Yet alongside this, our work points to a work-world which offers limited scope for commercial gain, can be boring and lonely, and is more often than not unpredictable and precarious – all qualities which we might more commonly equate with the conventional model of self-employment. More particularly, we argue that the economic and cultural position of retro trading is inherently unstable as it depends on high levels of self-exploitation and on eminently plagiarizable market conditions. While the barriers to entry are low in economic terms, in cultural terms commercial success is an altogether more problematic notion which requires constant innovation and distinction. Retro trading specifically, and the creative industries more broadly, seem to us to be a perfect example of where the economic and the cultural meet head-on and collide. And while in overtly cultural terms such sectors appear to offer a potentially more radical set of work options (where cultural attributes are prioritized and valorized), in narrowly economic terms such work practices seem neither alternative nor empowering. At the bottom line, retro traders, no matter how progressive, distinctive and pioneering their work plans may be, ultimately need to make enough money to stay in business and this, above all else, governs their market position, stability, permanence and positioning *vis-à-vis* the 'mainstream'. The all too common spatial outcome is to move up (into more mainstream commercial worlds), to move on (into more marginal and unpredictable spaces) or to move out of the retro trade altogether. The traditional model of small firms comes back to haunt us: transitory, precarious and economically marginal commercial success, it would seem, still relies on making profits through volume sales and low labour costs. And while such work may not be risky in the conventionally economic sense of the word, it is culturally an enormously risky undertaking, when what is at stake is one's taste, style, distinction and commercial credibility. The 'alternative' is, it would seem, but a temporary, imaginary space – as soon as it is definable, it is lost. The working practices of retro trading are hardly revolutionary; the spaces in which creative work occurs are even less so.

Secondly, our work is also significant in offering a new take on theorizations of creative work. Existing literature on creative work refers, we argue, primarily to those working at the less stable and more transient 'trading' end of the spectrum[15]. Such work focuses typically on new entrants to creative work and represents their employment experiences as both idealistic and static. And this focus is important in terms of what it has to say about work practices and (what it doesn't have to say) about career trajectories. Such work is predominantly cross-sectional in approach and offers a snapshot of creative workers at a particular moment in time and space. But what it does not do is to consider the ways in which workers' 'career' paths evolve and change, in many cases shifting from a transient market trading space to more formalized shops in more commercialized spaces. It is only by adopting a longitudinal approach, as we have done here, that we can really begin to interrogate the unstable and shifting boundaries between alternative and mainstream worlds (as imagined and practised). Only then, we argue, might it be possible to understand the shifting spatialities of creative work practices.

Finally, our work raises some serious questions about the ability of retro retailing to be transgressive, resistant, distinctive and critical, when it is so often incorporated into the mainstream, where it is reappropriated and subsumed, its meanings transformed and its power diluted. More specifically, our work raises some questions about the ability of retro retailing to survive, with high street chains threatening to plagiarize its imagination and creativity in a frantic copycat race. And what our work reveals is the disillusionment felt by retro retailers when business goes mainstream, thus making it much more difficult to position themselves as the creative, alternative other. The 'alternative', it seems, is constantly under attack from the nimble emulation of the 'mainstream'. And so it seems that the problem with notions of the alternative grounded in fashion discourse is that retro retailing is the very site of this creative dilemma. At the heart of the dilemma is the convergence between economic and symbolic capital. When creativity sells, its incorporation into the mainstream becomes seemingly inevitable, and in so doing it loses its creative potential, its alternative distinctiveness, its symbolic power.

Notes

This research was conducted with financial assistance from ESRC (R000222182). We would like to thank participants at the Manchester Institute for Popular Culture Conference at Manchester Metropolitan University (December 1999), at the Universities of Hull, Nottingham and NUS, Singapore, for stimulating discussions of earlier drafts of this paper.

1 The cultural industries, it is argued, are growing at almost twice the rate of the national economy, generate revenues of £50 billion per year, employ 982,000 people and generate an estimated value-added of some £25 billion and have export-earnings of £6.9 billion (Creative Industries Task Force, quoted in Leadbeater and Oakley, 1999: 10–11).

2 The creative industries sector is, not surprisingly, highly differentiated in both sectoral and organizational terms, and includes sole-trader operations working from home to large multinational organizations such as AOL-Time Warner. For our purposes here, we are restricting our analysis to small and micro businesses which operate

independently in organizational terms. We are not interested in large multinational corporations.

3 The key sectoral exception is the music industry which has been the focus of some important theoretical and empirical research in recent years. See, for example, Brindley (2000); Frith (1992); Hennion (1989); Leyshon, Matless and Revill (1999); Scott (1999).

4 The material on which this chapter draws was obtained through interviews with 21 retro retailers specializing in clothing and/or interiors and artefacts located in a range of designated 'alternative' trading sites in UK cities (Nottingham, Manchester, Bristol and London). The traders include some who have been in the business for many years and new entrants; include both men and women; and include some working on market stalls and some in more permanent trading situations working out of shops. These interviews were then supplemented by detailed observational work in key retro sites such as Manchester's Affleck's Palace, Nottingham's Baklash and London's Portobello Market in Notting Hill.

5 See Crewe, Brooks and Gregson (2003) for a fuller exposition of the unstable imaginative geographies of the alternative as revealed in retro retailer's discourses.

6 It is important to note here that we are not in any sense suggesting that our field notes are 'telling the truth' about the trade, nor are they being offered as a definitive version of what is going on here. Rather, we suggest that they offer another way of reading the spaces of retro trading, one which is of course deeply dependent on our own positionality as 'outsiders'. We were, in the majority of cases, 'unknowing' observers. And this positionality (as middle-class, white women academics) had some important impacts in terms of access to information. Two black interviewees in Notting Hill denied us access to information, arguing that we were in some way taking information from them to pursue our own careers while offering nothing back in return. This problem was compounded by the presence of a journalist from a fashion magazine who was also trying to secure access to information from retro shops in Notting Hill on the same day. One shop owner refused to talk to us, arguing that he'd already done three magazine interviews that day and was really more interested in making a sale than talking to journalists. This further endorses the problems facing creative quarters as media popularization threatens to undermine their distinctiveness. On other occasions our lack of knowledge about the spaces we were studying actually enhanced our access to information as we were seen as academics who were clearly 'outside' the industry and therefore did not pose any commercial threat.

7 Our frequent and repeated visits to a number of (closed) retro retail shops confirmed these flexible opening arrangements. The pattern of opening revealed little by way of routine or predictability. Fridays and

Saturdays were the only guaranteed days of opening for many traders.

8 The Arndale Centre is a large city-centre mall in Manchester that was relocated away from Oldham Street in Manchester's Northern Quarter to the city centre, thus leaving a large tract of semi-derelict and vacant land.

9 A Carwash is a 1970s theme disco club night where punters dress up in 1970s clothes and dance to 1970s and 1980s disco music.

10 It is worth noting here that certain formal retail spaces, such as Top Shop, are themselves adopting a more chaotic, rummagey spatial layout, emulating market-like spaces in order to re-inject some interest and fun back into predictable high street consumption.

11 Two large UK fashion chains.

12 Fannie's Attic, Daphne's Handbag, Helter Skelter and Full Circle in Nottingham; Freshman's in Sheffield; The Girl Can't Help It and One of a Kind in London are all examples of this tendency.

13 See Gregson, Brooks and Crewe (2000) for a fuller exposition of the commodification of retro.

14 Witness, for example, the media frenzy surrounding Portobello and Notting Hill following the release of the film *Notting Hill*, which has arguably pushed a formerly economically marginal yet culturally vibrant district towards mainstream, tourist-led conformity, as happened with Covent Garden in the early 1980s and Camden in the late 1980s. Recent media coverage of Hoxton and Spitalfields in the East End of London is counterposing Hoxton (hip and happening) against Notting Hill (tired and passé), again reinforcing the dangers of boundary definition and of oppositional arguments about the alternative (see, for example, Rickey, 2000).

15 We are referring here to work by, for example, Purvis (1999), who has looked at new, young pop fashion designers and McRobbie who has looked at market stallholders and young designers.

References

Allen, J. (1997) 'Reflexive modernisation: politics, tradition and aesthetics in the modern social order', book review in *Transactions of the Institute of British Geographers*, New Series, 22 (2): 263–4.

Allen, J. and Henry, N. (1997) 'Ulrich Beck's Risk Society at work: labour and employment in the contract service industries', *Transactions of the Institute of British Geographers*, New Series, 22 (2): 180–96.

Beck, U. (1992) *Risk Society: Towards a New Modernity*. London: Sage.

Beck, U., Giddens, A. and Lash, S. (1994) *Reflexive Modernisation*. Cambridge: Polity Press.

Bourdieu, P. (1993) *The Field of Cultural Production*. Cambridge: Polity Press.

Brindley, P. (2000) *New Musical Entrepreneurs*. London: Central Books.

Burnett, R. (1993) 'The popular music industry in transition', *Popular Music and Society*, 17: 87–114.
Burton-Jones, A. (1999) *Knowledge Capitalism: Business, Work and Learning in the New Economy*. Oxford: Oxford University Press.
Creative Industries Task Force (1998) Mapping Report. London: Department of Culture, Media and Sport/HMSO.
Crewe, L. (1996) 'Material culture: embedded firms, organisational networks and the local economic development of a fashion quarter', *Regional Studies*, 30 (3): 257–72.
Crewe, L. and Beaverstock, J. (1998) 'Fashioning the city: cultures of consumption in contemporary urban spaces', *Geoforum*, 29 (3): 287–308.
Crewe, L., Gregson, N. and Brooks, K. (2003) 'The discursivities of difference', *Journal of Consumer Culture*, 3 (1): 61–82.
DTI (1998) *Our Competitive Future*. Government White Paper, London: Department of Trade and Industry.
du Gay, P. (1996) *Consumption, Identity and Work*. London: Sage.
du Gay, P. (1997) *Production of Culture/Cultures of Production*. London: Sage/Open University.
Frith, S. (1992) 'The industrialisation of popular music', in J. Lull (ed.), *Popular Music and Communication*. Newbury Park, CA: Sage. pp. 49–74.
Frith, S. (1996) *Performing Rites: On the Value of Popular Music*. Cambridge, MA: Harvard University Press.
Garnham, N. (1990) *Capitalism and Communication: Global Culture and the Economics of Information*. London: Sage.
Grabher, G. (1993) *The Embedded Firm: The Socio-Economics of Industrial Networks*. London: Routledge.
Granovetter, M. (1991) 'The social construction of economic institutions', in A. Etzione and R. Eccles (eds), *Socio-economics: Towards a New Synthesis*. Armonk, NY: pp. 75–81.
Granovetter, M. (1992a) 'Problems of explanation in economic sociology', in N. Nohria and R. Eccles (eds), *Networks and Organisations: Forms and Action*. Boston, MA: Harvard Business School Press. pp. 25–56.
Granovetter, M. (1992b) 'Economic action and social structure: the problems of embeddedness', in M. Granovetter and R. Swedberg (eds), *The Sociology of Economic Life*. Oxford and Boulder, CO: Westview Press.
Gregson, N., Brooks, K. and Crewe, L. (2000) 'Bjorn again? Rethinking 70s revivalism through the reappropriation of 70s clothing', *Fashion Theory*, 5: 3–28.
Gregson, N. and Crewe, L. (2003) *Second Hand Worlds*. Oxford: Berg.
Hennion, A. (1989) 'An intermediary between production and consumption: the producer of popular music', *Science, Technology and Human Values*, 14: 400–23.
Hochschild, A. (1997) *The Time Bind: When Work Becomes Home and Home Becomes Work*. New York: Metropolitan Books.
Landry, C. (2000) *The Creative City*. London: Earthscan.
Leadbeater, C. (1999) *Living on Thin Air*. Harmondsworth: Penguin.

Leadbeater, C. and Oakley, K. (1999) *The Independents*. London: Demos.

Leyshon, A., Matless, D. and Revill, G. (1999) *The Place of Music*. New York: Guilford Press.

Lyotard, J.F. (1984) *The Postmodern Condition*. Minneapolis, MN: University of Minnesota Press.

McRobbie, A. (1989) *Zoot Suits and Second-Hand Dresses*. London: Macmillan.

McRobbie, A. (1998) *British Fashion Design: Rag Trade or Image Industry*. London: Routledge.

McRobbie, A. (1999) *In the Culture Society: Art, Fashion and Popular Music*. London: Routledge.

Negus, K. (1998) 'Cultural production and the corporation: musical genres and the strategic management of creativity in the US recording industry', *Media, Culture and Society*, 20: 359–79.

O'Connor, J. (1998) 'Popular culture, cultural intermediaries and urban regeneration', in T. Hall and P. Hubbard (eds), *The Entrepreneurial City*. Chichester: John Wiley. pp. 225–40.

O'Neill, P. and Gibson-Graham, J.K. (1999) 'Enterprise discourse and executive talk: stories that destabilize the company', *Transactions of the Institute of British Geographers*, New Series, 24: 11–22.

Pratt, A. (1997) 'The cultural industries production system: a case study of employment change in Britain, 1984–91', *Environment and Planning A*, 29: 1953–74.

Purvis, S. (1996) 'The interchangeable roles of the producer, consumer and cultural intermediary: the new pop fashion designer', in J. O'Connor and D. Wynne (eds), *From the Margins to the Centre*. Aldershot: Ashgate.

Rickey, M. (2000) 'Tat is where it's at', *Daily Telegraph*, 26 April: 12.

Scott, A. (1996) 'The craft, fashion and cultural products industries of Los Angeles: competitive dynamics and policy dilemmas in a multi-sectoral image production complex', *Annals of the AAG*, 86: 306–23.

Scott, A. (1999) 'The US recorded music industry: on the relations between organization, location and creativity in the cultural economy', *Environment and Planning A*, 31: 1965–84.

five

Alternative Work Spaces

Andrew Lincoln

A world of global markets and free-flowing capital poses fundamental challenges for workers and communities. While capital has used its mobility to exploit space, moving (and threatening to move) across borders in pursuit of higher returns, localities and nation states appear increasingly powerless to control the activities of multinational concerns (see Greider, 1997; Ohmae, 1990, 1995; although also see Dicken et al., 1997; Hirst and Thompson, 1996; Woods, 2000; Yeung, 1998). The ongoing erosion of Britain's manufacturing industry provides a stark example of how neo-liberal policies have made it easy for corporations to ride the waves of capital, shifting production to cheaper overseas locations. As Harvey highlights:

> Corporations ... have more power to command space, making individual places much more vulnerable to their whims. The global television set, the global car, become an everyday aspect of political-economic life. The closing down of production in one place and the opening up of production somewhere else became a familiar story – some large production operations have moved four or five times in the last twenty years. (Harvey, 2000: 64)

Abandonment and disinvestment by global capital has devastating implications for local labour markets, with the consequences for less-skilled workers being particularly severe. Indeed, studies have documented local struggles, illuminating the damaging impacts that job losses and poverty can have on a place (see Bluestone and Harrison, 1982; Clark, 1989; Scott and Storper, 1986). Such labour market problems lie at the heart of widening disparities in society, with workers and community members often left struggling to combat further financial and social exclusion (Lawless et al., 1998; Leyshon and Thrift, 1997; Martin, 1999).

While scrutinizing the globalization of economic activity and detailing the dire consequences for local labour markets, economic geographers have tended to neglect the actions and responses of workers and community members. In the context of global economic forces, local people have commonly been conceived as pawns in the system, unable to respond to the challenges posed by contemporary capitalism. As other chapters in this collection have highlighted, however, workers and community members remain capable of shaping the economic landscape through a range of alternative economic practices (see Aldridge et al., 2001; Fuller, 1998; Lee, 1996, 2000; Lincoln, 2000; Williams, 1996). Such community-based initiatives warrant increased attention from economic geographers as they prove there is scope for constructing alternative spaces in the face of an overwhelming capitalist conformity.

This chapter contributes to the growing corpus of work rethinking what counts as 'the economic' in economic geography, through focusing on the use of employee ownership as another local alternative. While the nature of contemporary employee ownership is diverse, including employee share schemes that convey only small proportions of ownership to workers, this chapter focuses on buy-outs that have enabled employees to acquire a majority stake in their business. It is argued that such buy-outs offer real possibilities for revolutionizing how companies are owned and managed. Employees acquiring a substantial ownership interest may provide the foundation from which a new approach to industrial relations can be established. Employee ownership also complements the range of possible alternatives detailed in this book, having the geographical potential to root capital in a locality, so sustaining jobs and local investment (Gates, 1998; Hirst, 1994, 1997; Logue et al., 1998; Wills, 1998). Taken together, these local initiatives provide the foundations for building a new, alternative economy that contrasts with the conventional priorities of 'shareholder' capitalism (see Hutton, 1999; Kelly et al., 1997).

The remainder of the chapter is divided into four parts. The first part begins with an examination of the historical roots of employee ownership. While a comprehensive review is beyond the chapter's remit, this section highlights how employee ownership has been embraced as an alternative option by a range of political movements and thinkers (see Lincoln, 1999, for an in-depth history). The second part provides details of more recent developments in employee ownership, with attention focusing upon the uneven growth of employee ownership within the United Kingdom. The third part then explores whether the opportunities offered by employee ownership are grasped in practice. Through an examination of the experiences of two firms, the chapter considers both the advantages and disadvantages of employees taking an ownership stake. Finally, the chapter concludes by assessing the potential of employee ownership to provide workers and communities with a sustainable future in the face of economic globalization.

THE ANTECEDENTS OF TODAY'S EMPLOYEE-OWNERS

The advocacy of employee ownership as an alternative option has a long history, being rooted in the Industrial Revolution and the formation of the working class. Growing discontent with the economic and social ills of industrial capitalism prompted Robert Owen to propose co-operative ideas as a counterpart to the capitalist mainstream. Owen's experience as a spinning-mill manager in Manchester brought him face to face with the dehumanizing and exploitative tendencies of capitalist production and its devastating impact on working people. Concerned by these conditions, Owen developed his own theories of an alternative system, which was based on co-operation rather than competition. In 1800 Owen began to put his co-operative ideas into practice by purchasing the cotton mills at New Lanark in Scotland. New Lanark was run in a way that combined efficient production with more humane working conditions, bringing Owen recognition for being one of the most philanthropic employers of the day. However, while seeking to help the working class, Owen's co-operative beliefs were firmly imposed 'from above' rather than led by the workers themselves. For instance, it was left to the middle class to govern New Lanark, while Owen placed emphasis on re-educating and re-moralizing the working class (see Hopkins, 1995; Thompson, 1964).[1]

His experiments at New Lanark led Owen to believe that capitalism could be replaced by a 'new moral world' comprising 'villages of co-operation' (see Owen, 1963). In 1824, he visited the United States and purchased over 30,000 acres of land in Indiana, which was used to establish the settlement of New Harmony. The following year, Owen created another self-supporting community at Orbiston in Scotland. However, both settlements rapidly dissolved due to a lack of capital and commitment from community members. While demonstrating that it was possible to achieve a co-operative alternative to prevailing free-market doctrines, these settlements were ultimately undermined by structural problems. Such difficulties are echoed in many of the alternative economic geographies explored in this book.

Despite the collapse of these communities, Owen played an important role in spreading co-operative sentiment among the working class during the 1820s. This sentiment was pulled together in a new working-class movement called Owenism (see Thompson, 1964). The movement echoed Owen's calls for the replacement of capitalism with a new co-operative world. For Owenites, their vision of an alternative society was rational, with capitalism viewed as a temporary phase, soon to be overthrown. As Pollard remarks:

> In their own eyes, the new vision of the Owenists was anything but Utopian. Capitalism, it seemed to them, could not last. ... Instead of enriching the population of this country, it impoverished it, instead of greater happiness, it created greater misery, poverty, and vice. All indications pointed to the transformation into a co-operative world of 'equal exchanges' without exploitation, without crises or unemployment, and without needless suffering. (Pollard, 1967: 106)

On his return from North America Owen became the head of the Owenite movement. Through this involvement, Owen became intimately involved in experiments with Equitable Labour Exchanges. These institutions allowed workers to exchange products using labour notes that recognized the hours of work that went into producing a commodity. From 1834 onwards, however, the influence of Owen and Owenism on working-class thought rapidly diminished, while other movements gained popular support.

While the Owenite movement declined during the 1830s, co-operative ideas were further developed by the Rochdale Pioneers in the 1840s. The Pioneers comprised a range of skilled workers, including weavers, wool sorters and shoemakers, who each contributed money towards establishing a co-operatively run shop. In opening their store at Toad Lane in 1844, the Pioneers attempted to provide their working-class customers with another way of obtaining basic necessities (Birchill, 1994; Gurney, 1996; Hopkins, 1995). Following the success of the Rochdale Pioneers' store, an increasing number of co-operative workshops were set up, resulting in the growth of the consumer co-operative movement.

Co-operative principles were also put forward as an alternative by the Christian socialist movement. The Christian socialists largely comprised middle-class members of the Church of England who were troubled by the conditions endured by the working class. From 1848 onwards, this religious movement, led by J.F.D. Maurice, Charles Kingsley, John Ludlow and Edward Neale, embraced co-operative ideals with the aim of establishing a new model for society. Inspired by the experiences of the Rochdale Pioneers, the Christian socialists fostered the development of co-operative workshops and distributive societies. They also formed a 'Society for Promoting Working Men's Associations' (see Backstrom, 1974; Cole, 1944; Webb, 1906). However, the Christian socialist movement ended abruptly in 1854 following the collapse of several of their co-operatives. Such problems prompted the movement's leaders to wind up the Society and concentrate instead on educating the working class.

While the Christian socialist movement was relatively short-lived, other European political movements continued to consider employee ownership as an alternative to capitalism, with anarchist and Marxist debates being particularly influential. While sharing a commitment to overthrow capitalism, each of these political movements accorded a different priority to co-operatives in the alternative system.

Co-operative ideas were supported by Pierre Joseph Proudhon, an influential figure in the anarchist movement. In place of the existing capitalist system, Proudhon advocated an anarchist alternative, devoid of property and authority. This alternative vision was to be founded on principles of mutuality, with working people governing themselves, both in the workplace and wider society. In an attempt to put these principles into practice, Proudhon called for the establishment of federated communes and worker co-operatives. He was also involved in the creation of co-operative credit banks, where money could be borrowed without interest. This mutual banking system allowed workers to exchange products by means of labour cheques. In a system similar to Owen's Equitable Labour Exchanges, cheques were used to enable workers to exchange commodities (see Edwards, 1970; Rothschild and Whitt, 1989). However, when Proudhon was found guilty of attacking the Government, the bank's operations were suspended, never to re-open (Hyams, 1979).[2]

While occupying an important position in anarchist philosophy, co-operatives also featured in Marxist thought. For Karl Marx, co-operative experiments highlighted that it was possible to achieve production without the need for a class of capitalists. In his 1864 inaugural address to the International Working Men's Association, Marx recognized the value of 'these great social experiments':

> [T]here was in store a still greater victory of the political economy of labour over the political economy of property. We speak of the cooperative movement, especially the cooperative factories raised by the unassisted efforts of a few bold 'hands'. The value of these great social experiments cannot be overrated. By deed, instead of by argument, they have shown that production on a large scale, and in accord with the

> behests of modern science, may be carried on without the existence of a class of masters employing a class of hands; that to bear fruit, the means of labour need not be monopolized as a means of domination over, and of extortion against, the labouring man himself. (Marx, 1983: 363)

Recognizing the contribution of co-operatives, Marx viewed them solely as a lesson in a wider struggle to free the masses and abolish class relations. For Marx, worker ownership did not have the revolutionary potential to offer an alternative to capitalism. Instead, the transition to a classless society was to be achieved through a revolution in economic and political relations. Nevertheless, once this wider revolution had occurred, Marx believed it would be possible to expand co-operatives 'to national dimensions ... by national means'. Therefore, while co-operatives were never central to Marx's vision of a classless society, he recognized their potential to play an important role in a communist system.

While embraced by a range of political movements and thinkers, commitments to worker ownership were increasingly superseded by new priorities as industrialization advanced. In Britain, where co-operative ideas flourished during the early years of the Industrial Revolution, workers turned to trade unions to represent their interests. Instead of pursuing radical alternatives, these unions increasingly accepted the capitalist system, focusing on the extension of public ownership and collective bargaining. While public ownership assumed a central position in the development of the British working class, co-operative ideas continued to recur at the fringes of the movement. Indeed, Robert Oakeshott (1990: 51) identifies a 'motley array of splinter groups and divergent factions' committed to employee ownership, operating within the socialist mainstream. In particular, guild socialist thinkers continued to call for the rejection of capitalist or state control of industry during the early years of the twentieth century (Cole, 1944). Influenced by French Syndicalism, guild socialists advocated workers' control of industry and the restoration of the medieval guild system. The direct participation of workers in production was viewed as the first step towards the democratization of society (see Carpenter, 1922; Cole, 1917). However, their alternative vision was never realized and, ultimately, guild socialist ideas remained consigned to the edges of the working-class movement, declining in influence as the 1920s progressed.

Worker ownership was also revived as an alternative option during periodic downturns in the British economy. While post-war Labour governments were committed to nationalization, employee ownership was given a new lease of life in the 1970s. In the context of economic stagnation and rationalization, the Labour government advocated the extension of industrial democracy and workers' control as the solution to Britain's business problems. As part of this strategy, Tony Benn, the Secretary of State for Industry, supported employee ownership as a last resort at three failing firms. Co-operatives were established at Meriden motorcycles near Coventry, the Glasgow-based Scottish Daily News, and Kirkby Manufacturing on the outskirts of Liverpool. These ventures all proved to be short-lived, however, with low levels of profit bringing an abrupt end to all three co-operatives (Bradley and Gelb, 1983; Coates, 1976; Fleet, 1976). While the failure of the 'Benn experiments' served to further discredit co-operatives, the subsequent two decades have witnessed the re-emergence of employee ownership ideas.

CONTEMPORARY EMPLOYEE OWNERSHIP DEVELOPMENTS

Having been put forward as an alternative over time, there has recently been a resurgence of employee ownership sentiment in the United Kingdom. This began during a period of Conservative rule that fundamentally altered Britain's political and economic landscape. In a climate promoting free market capitalism and privatization, employee ownership was once again advocated as a valuable alternative to conventional priorities. In particular, when privatization was proposed, employee ownership was considered as another option, allowing workers to defend their jobs and communities.

During this period of Conservative hegemony, the left was forced to rethink its priorities, including its attitude towards employee ownership. Where privatization was inevitable, the left increasingly pursued employee ownership in preference to management buy-outs or take-over by private enterprise. This was the case in the bus industry, where several Labour-controlled local authorities viewed employee buy-outs as the best option for protecting jobs, working conditions and services (Pendleton et al., 1996; Wright et al., 1992).[3] Following the election of a Labour government in 1997, employee ownership has also received further support. As part of its commitment to workplace partnership, the government has attempted to encourage employee ownership as a 'third way' option between public and private ownership (see Amin, 1996; Cressey, 1999; Froud et al., 1996).

The growth of employee ownership in the UK has remained very slow, however, and it is estimated that employee take-overs account for only 1 per cent of buy-outs (see Monks, 1999).[4] British employee ownership developments have also been geographically uneven, with buy-out feasibility and the ability of workers to acquire an ownership stake varying across space.[5] Such variations reflect distinct industrial cultures and place-based allegencies that exist in particular communities (see Massey, 1984). In this respect, employee ownership cannot be divorced from its geographical context, rooted in local social relations and labour market characteristics.

The importance of local social relations as the foundation of economic alternatives is shown in the remainder of this chapter by examining the employee ownership experiences at a British coal mine and bus company. The contrasting fortunes of these two cases illustrate how a range of factors must be in place for employee ownership to fulfil its potential. Both case studies highlight how workers are better equipped to realize the possibilities of employee ownership in some places than in others.[6]

BRINGING HOPE TO A COMMUNITY? THE EMPLOYEE BUY-OUT AT TOWER COLLIERY

With the British coal industry boasting 958 deep coal mines and 718,000 workers when nationalized by the Labour government in 1947, the extent of mine closures across the UK has been staggering, with only 32 collieries and 9,000 employees remaining by the mid-1990s (see Parry et al., 1997). As Wales's last deep mine in a region once employing 270,000 miners, the announcement that Tower colliery was to be closed was a bitter blow for workers and community members alike. Despite generating profits of £28 million in the preceding three years, the 'uneconomic' colliery was selected for shutdown as part of the Conservative government's pit closure programme. The decision was devastating for residents of Hirwaun and the surrounding Cynon Valley, an area already suffering from high levels of unemployment. While the traditionally militant workforce mounted protest marches and a 48-hour sit-in to save the pit, their efforts seemed to no avail when the colliery was closed in April 1994. The closure ended 235 years of history, with severe implications for the local economy which had grown heavily dependent on jobs provided at the pit.

Unlike so many other shutdowns, however, the closure at Tower colliery proved not to be terminal. Determined to prevent permanent closure or a rumoured management buy-out, local union officials proposed an employee take-over as an alternative strategy for preserving members' jobs. Supported by their local community, the former miners embarked on an eight-month struggle for ownership. Following a series of complex negotiations with Price-Waterhouse, Barclays Bank and the Department for Trade and Industry, workers were eventually able to buy back their colliery in December 1994. To become owners, each employee invested £8,000, raising £1.92 million to make Tower the only 100 per cent employee-owned pit in Europe.

The success of the Tower buy-out is rooted in distinctive social relations, with the industrial culture of the region reflecting a long history of trade union organization. Workers' shared histories and traditions of resistance proved an important source of strength during the buy-out and were also crucial in subsequently reconfiguring workplace relations. While adversarial industrial relations had traditionally plagued the colliery under British Coal, the transition to employee ownership was used as an opportunity to introduce a new approach to running the business, based on partnership. At the heart of this new vision was the aim of employing as many workers as possible from the surrounding community, enabling local people to enjoy a better standard of living through decent terms and conditions. To achieve such principles, there was recognition that Tower must operate in the marketplace in order to survive, as the Personnel Director (and former NUM Lodge Secretary) revealed:

> [W]hat we were doing here wasn't joining the capitalists, we were forming a company, a predominantly socialist company, socialist beliefs, having to work in a capitalist world ... we have got to work within the rules of capitalism and trade with capitalist countries. We have always got to make a profit. ... So we just deal with capitalists and I think the integrity and honesty that we have brought into the capitalist system has been invigorating. When other capitalists deal with us, they are surprised at our integrity and honesty to such an extent that we get an advantage. (Interview, July 1998)

To achieve these objectives, a new management structure was imposed, with two worker directors being appointed to run the firm jointly with three executive directors. While continuing to be represented by the National Union of Mineworkers (NUM), employees were also encouraged to participate in business decisions through joint union–management consultation meetings and quarterly shareholder meetings (see Parry et al., 1997).

Despite the substantial costs associated with re-opening the colliery, Tower generated a £3.6 million pre-tax profit at the end of its first year as an employee-owned business. Such success has enabled a range of wider social objectives to be met. While sustaining their own jobs, Tower workers have also generated further job opportunities at the pit, raising employment prospects for local people. As the Chair of Tower NUM described:

> [W]e have done what we said we would do in creating employment and so forth. I mean you might see young people walking around here now. We have got about twenty young boys of sixteen, seventeen, eighteen and nineteen that actually came off the streets. If we hadn't taken them off the streets, they could have been on drugs or anything. So it is giving something back to the community and I take great pride in what we have done. (Interview, July 1998)

In addition to generating new employment opportunities, the employee-owned colliery has had positive implications for other local firms, contributing to regional development. As the Chair of Tower NUM went on to reveal:

> [A]ll of the profits from our colliery are actually kept within the valleys. It doesn't go to London or anywhere else, it is actually spent in the valleys. As long as the price is okay, we try to deal with local businesses as much as possible to keep other people in work. (Interview, July 1998)

Recognizing the backing they received during their buy-out, workers have also attempted to support a number of local projects that meet community needs. This has included providing help to the unemployed, supporting drugs and alcohol abuse initiatives and putting money into a disabled riding school.

Despite successfully pursuing a range of alternative priorities, Tower continues to experience occasional tensions that can strain the shared traditions of the workforce. The potential for conflict has continued to exist among different sections of the workforce, with pre-existing pay grade differentials between surface and coalface workers being particularly divisive. In an attempt to resolve such tensions, a pay committee was established to review the wage structure within the company and a 4.5 per cent pay rise was later implemented. These basic pay rates compare favourably with those paid by privately-owned RJB Mining, while Tower has also introduced better benefits than previously existed under British Coal (see Parry et al., 1997).

Such improvements have not prevented periodic disagreements between groups of the workforce and management, culminating in 100 workers going on strike for a day in April 2000 when managers ordered two employees to change their work schedule. In such circumstances, maintaining employee ownership has taken a lot of effort, requiring both union and management representatives to work together to resolve difficulties as they arise. The need to manage conflicts and keep workers pulling in the same direction was recognized by the Personnel Director:

> This company is such a company now that the workforce fear no threats, they are satisfied with their lives ... so now you have got to try and find them an enemy outside the company, keep them loyal to each other. Instead of us becoming the enemy as directors or management, let's find an enemy outside so that we all pull together to face that enemy, rather than look at each other. (Interview, July 1998)

The lack of alternative employment opportunities also provides a strong unifying influence within the company. Having collectively struggled to buy back their colliery, workers remain committed to resolving disagreements, making their employee-owned business a success. While it remains to be seen whether Tower will continue to be sustainable in the long term, it has been operated profitably as an employee-owned business, so bringing hope to the local community.

STAKE OR MISTAKE? EMPLOYEE OWNERSHIP AT BUSCO[7]

BusCo is a bus company located in a city in the South of England. Having been municipally owned for over 100 years, the employee buy-out of the business was prompted by Conservative attempts to deregulate and privatize public transport (see Hutton, 1997). Expecting further government legislation requiring local authorities to dispose of their bus companies to private bidders, the Labour-controlled City Council decided to investigate the possibility of employees owning the firm. With the Council committed to providing a community bus service, worker ownership was viewed as the best strategy for retaining a local interest in the business, benefiting both employees and the wider community. While actively promoted by the local Council, employees also supported the buy-out as a more favourable option to being taken over by an outside company. Being sold to the highest bidder was seen as potentially having very damaging consequences, leading to redundancies, as well as reductions in wages and working conditions. To prevent this from occurring, 330 employees joined together, each investing £1,200 to become the owners of their firm in December 1993.[8]

The BusCo buy-out was plagued with problems from the start, however, with local social relations again exerting influence on events.

In particular, a lack of shared traditions and rivalries between different unions operating in the company combined to undermine the buy-out. This resulted in workers' interests not being best represented during early negotiations, with union officials focusing on wages and working conditions rather than on the structure of the deal itself. Local union representatives did not even seek independent legal and financial advice about the buy-out. As a result, the buy-out was predominantly shaped by senior managers and their professional advisors, with damaging consequences for how the business was subsequently run.

The transition to employee ownership at BusCo was accompanied by the imposition of a new management structure, with traditional collective bargaining institutions being supplemented by the introduction of worker directors and employee trustees. Despite these new forms of representation, BusCo managers made no attempts to reconfigure traditional 'them and us' relationships, failing to establish an alternative approach to the business. Instead, management pursued conventional Anglo-American business practices in a quest for profit and shareholder value. For BusCo's Managing Director, employees were simply shareholders, investing in the buy-out for the same reason that people invest in Marks & Spencer, that is, to realize a profit:

> [I]f you go out ... and become a shareholder in Marks & Spencer, the reason you have made that decision is not for some altruistic view that you want to see better clothes in Marks & Spencer or you want to see your favourite food being displayed on the shelves. You want to make a profit out of that investment. (Interview, September 1997)

Moreover, senior managers remained opposed to involving employees in the running of the business, asserting 'management's right to manage'. While the Managing Director recognized that the workforce had offered suggestions about how the firm should operate, he argued it was the management's prerogative to make decisions regarding the business:

> [T]he workforce have their own ideas as to how the business wants to operate and the way they would like to operate it, but they have no responsibility. If some of their ideas, although fashionable as it may seem to a bus driver, but to a professional manager are totally irresponsible – what is the director to do? Is the director to follow the less than practical suggestions of the shareholder or is he there to follow the Companies Act and to operate with his fiduciary duties? (Interview, September 1997)

In this quest for profit, management decisions were imposed on the workforce rather than explained, resulting in further animosity. Such an approach did not meet workers' expectations of employee ownership and it was felt that the managers had forgotten that the workforce had also invested in the buy-out, as the following excerpt from an employee group interview illustrates:

> Employee 1: In all my life, this is the worst company I have ever worked for. The worst management I have ever worked for. It might be employee-owned but the employees don't get any say in it.
> Employee 2: I think employee 'con' would be a more accurate word.
> Employee 1: The trouble is the bloke up there [the Manager Director] thinks he bought it with his money. ... It was all our £1,200 that went into it. But afterwards he didn't want to know. (Employee group interview, April 1997)

While certain workers campaigned to change management priorities, others increasingly considered the option of being taken over by an outside company. Demoralized with employee ownership as practised in their firm, workers were lured by talk of the financial rewards available for selling their stake. These rumours were backed up by stories about bus workers from other employee-owned firms who had sold their stake in return for large sums of money (see Barrie, 1998; Ford and Jones, 1998; Gresser, 1997). Despite attempts at reasoning with these employees about why a sale was ultimately not in their best interests, the arguments against take-over fell on deaf ears. As the local Branch Secretary explained, there were good reasons for employees wanting to sell:

> A lot of people are just fed up with the situation. ... The things that they were promised haven't happened and therefore a lot of people are saying: 'Well we might as well work for another company. If we are going to get a couple of quid in the process, you know, that's fine.' I don't blame those people for thinking like that because they have been badly let down by the employee ownership theory within this company. They have been badly let down. (Interview, April 1997)

With the BusCo board recommending a take-over deal, the workforce were balloted and over 90 per cent voted in favour of the sale, realizing a return of £15,000 on their initial £1,200 stake. As one employee commented: 'I've got my pen ready, show me where to sign. My pen is burning a hole in my pocket.' Nevertheless, there was still resentment from some workers that the opportunities of employee ownership had been wasted. As one worker stated: 'I feel a bit like a Judas personally 'cos I'm taking my twenty pieces of gold and watching the future generations being stitched up by it' (Employee interview, April 1997).

In its three years as an employee-owned company, BusCo was operated as a profitable business, fulfilling management objectives. However, had a new corporate culture been established and employee participation encouraged, arguably the firm could have been even more successful. Unfortunately, employee ownership at BusCo proved only to be a transitionary stage, crumbling under worker disillusionment and growing market pressures. The only positive feature of this brief interlude of employee ownership was that the entire workforce was able to share in the wealth of the business when it was sold.

THE ADVANTAGES AND DISADVANTAGES OF EMPLOYEE OWNERSHIP AS AN ALTERNATIVE OPTION

The contrasting cases of Tower colliery and BusCo illustrate the possibilities and problems of employee ownership as an alternative strategy. While highlighting the advantages of workers taking an ownership stake, the chapter has simultaneously sounded a note of caution. By exploring how buy-outs work in practice, it has been shown that a number of challenges remain to be conquered for employee ownership to be a success. Both cases illustrate the importance of local circumstances and, in particular, the institutional shaping of local social relations in determining employee ownership's potential to succeed as an alternative economic space.

At Tower colliery, the prevailing industrial culture and shared histories that had for so long united workers against their employer were crucial in securing the transition to employee ownership. While enabling workers to become owners, these locally constituted labour traditions also made it possible for workplace relations to be reshaped following the buy-out. Through the introduction of democratic governance structures, employees were able to participate in decision-making, with positive implications for business performance. Despite experiencing periodic disagreements and remaining market-dependent, the colliery has provided good-quality local jobs, as well as offering new employment opportunities.

While Tower colliery highlights buy-out possibilities, the BusCo case once again illustrates the important nature of social relations in the creation of alternative economic spaces. Many of the problems experienced by BusCo workers are grounded in the shaping of local social relations through the institution of a local state as well as in local characteristics and events, including access to and acceptance of the norms of finance capital. BusCo's problems stem from the way the buy-out deal was originally structured. In contrast to Tower, the BusCo buy-out was initiated 'from above' by the local Council, rather than led by the workers themselves. In particular, BusCo workers lacked the shared histories and long-standing traditions that served to unite Tower workers. While some new governance structures were introduced following the buy-out, employee representatives became absorbed into the quest for profits, and were viewed as 'company spies' by certain sections of the workforce. With managers not interested in adopting a new corporate culture, the company continued to be run as a conventional capitalist business, resulting in employee ownership ultimately being only a temporary phase. The BusCo case thus highlights the importance of workers being involved both during and following the transition to employee ownership. While 'owners on paper', BusCo employees continued to have minimal input in their business and employee ownership was not seized as a real tool for change.

CONCLUSIONS

This chapter has highlighted how economic geographers have traditionally tended to neglect the actions of workers and community members in contemporary accounts of the global economy. While detailing the challenges posed by globalization and multinational capital flows, the chapter has argued that employees and communities remain capable of forging their own responses. In so doing, this chapter has contributed to the growing body of geographical scholarship scrutinizing local economic alternatives. Recognizing the valuable contribution made by these studies, the research presented in this chapter has added a new dimension by analysing the ability of employee ownership to aid community economic development. Through negotiating an ownership stake in their workplace and exercising their governance rights, it has been argued that employees potentially have the opportunity to construct alternative spaces, shaping the economic landscape on their own terms. Like other local alternatives, however, employee ownership may not be applicable in every situation, with local circumstances influencing the extent to which workers can seize the initiative.

As British capitalism continues to prioritize short-term shareholder interests, employee ownership remains worthy of encouragement as one of a range of possible alternative options. At present, however, there remains a need for such local initiatives to be moved up the political agenda. The British government, particularly, has a key role to play in further developing the employee ownership landscape, through highlighting successful cases and learning from negative experiences. The recent creation of Social Enterprise London (SEL) marks a step in the right direction. This new regional organization is committed to promoting social enterprises in London and is ideally placed to encourage employee buy-outs. There remains a need, however, for further regional institutions to promote employee ownership ideas and publicize examples of best practice (see Wills and Lincoln, 1999).

At a time when scholars are rethinking what constitutes 'the economic' in economic geography, the ability of employee ownership to operate at the margins of the capitalist system warrants further attention. With communities increasingly vulnerable in today's global age, taking an ownership stake may allow workers to fix capital in their locality, sustaining local jobs. While harnessing capital for local communities, employee ownership may also offer opportunities for introducing alternative business practices that combine improved workplace relations with wider social aims. When used in this way, employee ownership potentially offers workers an alternative economic space for resisting globalization.

Notes

My first debt of gratitude is to Tower and BusCo workers who took part in this research. I would also like to thank Jane Wills, Adam Tickell and Jamie Peck for useful feedback that helped me to refine some of the ideas presented in this chapter. Finally, I am extremely grateful to Roger Lee, Andrew Leyshon and Colin Williams for their support and encouragement as editors. Their useful comments on an earlier draft of this chapter are particularly appreciated.

1 Following Owen, other philanthropist employers established their own model communities in an attempt to combine better working conditions with improved living standards. In 1851, the community of Saltaire was built by Sir Titus Salt, while Port Sunlight was created by the Lever Brothers in 1887. Other model settlements include, Peacehaven in Sussex, Portmeirion in Wales and Bournville, established by George Cadbury in Birmingham (Thornley, 1981). Some philanthropic employers, including John Spedan Lewis, Ernest Bader

and Philip Baxendale, even handed over flourishing businesses to their workforces (Bradley and Gelb, 1986).

2 Anarchist geographer Peter Kropotkin also became interested in co-operative issues following the publication of Darwin's *Origin of Species* (1859) and the growth of Social Darwinism. Questioning the struggle for existence proposed in Darwin's theory of evolution, Kropotkin's *Mutual Aid* was published in 1902, emphasizing the importance of within-species co-operation, rather than competition. Recognizing the historical importance of co-operation in the evolutionary process, Kropotkin argued that mutual aid would form the basis of a new society, comprising a multitude of associations (Kropotkin, 1972; Read, 1942; Shatz, 1995).

3 From the mid-1990s onwards, however, many of these employee-owned firms were taken over by larger operators, such as Stagecoach and Firstbus (see Hutton, 1997).

4 While employee-owned businesses remain limited in the context of British capitalism, such ventures continue to provide important job opportunities for the people they employ.

5 In addition, the development of employee ownership varies between nations. For instance, there is tremendous dynamism behind the employee ownership movement in the USA, with over 3,000 majority employee-owned businesses, employing more than 1.5 million people. These businesses comprise small and medium-sized enterprises, as well as larger corporations, such as United Airlines (see Wills and Lincoln, 1999 for further details on North American developments). The United States has also witnessed some of the worst employee ownership abuses. This has included employee ownership being introduced in 'unviable' situations, with workers buying into their business, only to see it subsequently shut down (see Prude, 1984; Quarter, 1989; Russell, 1984; also see Waddington et al., 1998 on the case of Monktonhall colliery in Scotland).

6 The interview material presented in this chapter is drawn from meetings held with different members of the workforce from the case study companies during 1997 and 1998.

7 BusCo's fictitious name was chosen to protect the identities of the research participants.

8 The remainder of the purchase price was covered by a combination of council and bank loans.

References

Aldridge, T., Tooke, J., Lee, R., Leyshon, A., Thrift, N. and Williams, C. (2001) 'Recasting work: the example of Local Exchange Trading Schemes', *Work, Employment and Society*, 15: 565–79.

Amin, A. (1996) 'Beyond associative democracy', *New Political Economy*, 1 (3): 309–33.

Backstrom, P.N. (1974) *Christian Socialism and Cooperation in Victorian England*. London: Croom Helm.
Barrie, C. (1998) 'Bus takeover deal puts workers in line for £12,000 windfalls', *The Guardian*, 2 June 1998, p. 23.
Birchill, J. (1994) *Co-op: The People's Business*. Manchester: Manchester University Press.
Bluestone, B. and Harrison, B. (1982) *The Deindustrialisation of America*. New York: Basic Books.
Bradley, K. and Gelb, A. (1983) *Worker Capitalism: The New Industrial Relations*. London: Heinemann Educational Books.
Bradley, K. and Gelb, A. (1986) *Share Ownership for Employees*. London: Public Policy Centre.
Carpenter, N. (1922) *Guild Socialism: An Historical and Critical Analysis*. London: D. Appleton and Company.
Clark, G.L. (1989) *Unions and Communities Under Siege: American Communities and the Crisis of Organized Labor*. Cambridge: Cambridge University Press.
Coates, K. (ed.) (1976) *The New Worker Co-operatives*. Nottingham: Spokesman Books.
Cole, G.D.H. (1917) *Self Government in Industry*. London: Bell and Sons.
Cole, G.D.H. (1944) *A Century of Co-operation*. London: George Allen and Unwin.
Cressey, P. (1999) 'New Labour and employment, training and employee relations', in M. Powell (ed.), *New Labour, New Welfare State? The 'Third Way' in British Social Policy*. Bristol: The Policy Press.
Dicken, P., Peck, J. and Tickell, A. (1997) 'Unpacking the global', in R. Lee and J. Wills (eds), *Geographies of Economies*. London: Arnold.
Edwards, S. (ed.) (1970) *Selected Writings of Pierre-Joseph Proudhon*. London: Macmillan.
Fleet, K. (1976) 'Triumph Meriden', in K. Coates (ed.), *The New Worker Co-operatives*. Nottingham: Spokesman Books. pp. 88–108.
Ford, J. and Jones, S. (1998) 'Busmen find share ownership may not be a ticket to ride', *The Financial Times*, 24/25 October 1998, p. 8.
Froud, J., Haslam, C., Johal, S., Shaoul, J. and Williams, K. (1996) 'Stakeholder economy? From utility privatisation to New Labour', *Capital and Class*, 60: 119–34.
Fuller, D. (1998) 'Credit union development: financial inclusion and exclusion', *Geoforum*, 29 (2): 145–57.
Gates, J. (1998) *The Ownership Solution: Toward a Shared Capitalism for the Twenty-first Century*. Reading, MA: Addison-Wesley.
Greider, W. (1997) *One World, Ready or Not: The Manic Logic of Global Capitalism*. New York: Simon & Schuster.
Gresser, C. (1997) 'Metroline employees to net £9,000', *The Financial Times*, 3 June 1997, p. 23.
Gurney, P. (1996) *Co-operative Culture and the Politics of Consumption in England, 1870–1930*. Manchester: Manchester University Press.
Harvey, D. (2000) *Spaces of Hope*. Edinburgh: Edinburgh University Press.

Hirst, P. (1994) *Associative Democracy: New Forms of Economic and Social Governance*. Cambridge: Polity Press.
Hirst, P. (1997) *From Statism to Pluralism*. London: UCL Press.
Hirst, P. and Thompson, G. (1996) *Globalization in Question: The International Economy and the Possibilities of Governance*. Cambridge: Polity Press.
Hopkins, E. (1995) *Working-class Self-help in Nineteenth-century England: Responses to Industrialisation*. London: UCL Press.
Hutton, W. (1997) *The State to Come*. London: Vintage.
Hutton, W. (1999) *The Stakeholding Society: Writings on Politics and Economics*. Cambridge: Polity Press.
Hyams, E. (1979) *Pierre-Joseph Proudhon: His Revolutionary Life, Mind and Works*. London: John Murray.
Kelly, G., Kelly, D. and Gamble, A. (eds) (1997) *Stakeholder Capitalism*. London: Macmillan.
Kropotkin, P. (1972) *Mutual Aid: A Factor of Evolution*. London: Penguin.
Lawless, P., Martin, R. and Hardy, S. (eds) (1998) *Unemployment and Social Exclusion: Landscapes of Labour Inequality*. London: Jessica Kingsley.
Lee, R. (1996) 'Moral money? LETS and the social construction of economic geographies in South East England', *Environment and Planning A*, 28: 1377–94.
Lee, R. (2000) 'Radical and postmodern? Power, social relations, and regimes of truth in the social construction of alternative economic geographies', *Environment and Planning A*, 32: 991–1009.
Lee, R. and Wills, J. (eds) (1997) *Geographies of Economies*. London: Arnold.
Leyshon, A. and Thrift, N. (1997) *Money/Space: Geographies of Monetary Transformation*. London: Routledge.
Lincoln, A. (1999) 'Revolution at work? Employee buy-outs in a stakeholder society'. Unpublished PhD thesis, University of Southampton.
Lincoln, A. (2000) 'Working for regional development: the case of the Canadian labour-sponsored funds', *Regional Studies*, 34 (8): 727–37.
Logue, J., Glass, R., Patton, W., Teodosio, A. and Thomas, K. (1998) *Participatory Employee Ownership: How It Works*. Pittsburgh, PA: The Worker-Ownership Institute.
Martin, R. (ed.) (1999) *Money and the Space Economy*. Chichester: John Wiley.
Marx, K. (1983) 'Inaugural address and provisional rules of the International Working Men's Association', in E. Kamenka (ed.), *The Portable Karl Marx*. New York: Penguin. pp. 355–68.
Massey, D. (1984) *Spatial Divisions of Labour: Social Structures and the Geography of Production*. London: Macmillan.
Monks, J. (1999) 'The "own" in ownership', *The Times Higher*, 26 March, p. 31.
Oakeshott, R. (1990) *The Case for Workers' Co-ops* (second edition). London: Macmillan.
Ohmae, K. (1990) *The Borderless World*. London: Collins.

Ohmae, K. (1995) *The End of the Nation State*. New York: The Free Press.

Owen, R. (1963) *A New View of Society and Other Writings*. London: Dent and Sons.

Parry, D., Waddington, D. and Critcher, C. (1997) 'Industrial relations in the privatized mining industry', *British Journal of Industrial Relations*, 35 (2): 173–96.

Peck, J. (1996) *Work-place: The Social Regulation of Labor Markets*. New York: Guilford.

Pendleton, A., McDonald, J., Robinson, A. and Wilson, N. (1996) 'Employee participation and corporate governance in employee-owned firms', *Work, Employment and Society*, 10 (2): 205–26.

Pollard, S. (1967) 'Nineteenth-century co-operation: from community building to shopkeeping', in A. Briggs and J. Saville (eds), *Essays in Labour History*. London: Macmillan. pp. 74–112.

Proudhon, P.J. (1970) 'Mutualism', in S. Edwards (ed.), *Selected Writings of Pierre-Joseph Proudhon*. London: Macmillan. pp. 56–70.

Prude, J. (1984) 'ESOP's fable: how workers bought a steel mill in Weirton, West Virginia and what good it did them', *Socialist Review*, 78: 26–60.

Quarter, J. (1989) 'Worker ownership: one movement or many?', in J. Quarter and G. Melnyk (eds) *Partners in Enterprise: The Worker Ownership Phenomenon*. Quebec: Black Rose Books. pp. 1–32.

Read, H. (ed.) (1942) *Kropotkin: Selections from His Writings*. London: Freedom Press.

Rothschild, J. and Whitt, J.A. (1989) *The Cooperative Workplace: Potentials and Dilemmas of Organizational Democracy and Participation*. Cambridge: Cambridge University Press.

Russell, R. (1984) 'Using ownership to control: making workers owners in the contemporary United States', *Politics and Society*, 13: 253–94.

Sawers, L. and Tabb, W.K. (eds) (1984) *Sunbelt/Snowbelt: Urban Development and Regional Restructuring*. Oxford: Oxford University Press.

Scott, A. and Storper, M. (eds) (1986) *Production, Work, Territory: The Geographical Anatomy of Contemporary Capitalism*. Boston, MA: Allen and Unwin.

Shatz, M. (ed.) (1995) *Peter Kropotkin: The Conquest of Bread and Other Writings*. Cambridge: Cambridge University Press.

Thompson, E.P. (1964) *The Making of the English Working Class*. London: Victor Gollancz.

Thornley, J. (1981) *Workers' Co-operatives: Jobs and Dreams*. London: Heinemann Educational Books.

Waddington, D., Parry, D. and Critcher, C. (1998) 'Keeping the red flag flying? A comparative study of two worker takeovers in the British deep coalmining industry, 1992–1997', *Work, Employment and Society*, 12 (2): 317–49.

Webb, C. (1906) *Industrial Co-operation: The Story of a Peaceful Revolution*. Manchester: The Co-operative Union.

Williams, C.C. (1996) 'Local exchange trading systems in the United Kingdom: a new source of work and credit for the poor and unemployed', *Environment and Planning A*, 28 (8): 1395–415.

Wills, J. (1998) 'A stake in place? The geography of employee ownership and its implications for a stakeholding society', *Transactions of the Institute of British Geographers*, 23 (1): 79–94.

Wills, J. and Lincoln, A. (1999) 'Filling the vacuum in "new" management practice? Lessons from American employee-owned firms', *Environment and Planning A*, 31 (8): 1497–512.

Woods, N. (ed.) (2000) *The Political Economy of Globalization*. London: Macmillan.

Wright, M., Dobson, P., Thompson, S. and Robbie, K. (1992) 'How well does privatisation achieve Government objectives? The case of the bus buy-outs', *CMBOR Occasional Paper 38*. Nottingham: University of Nottingham.

Yeung, H. W. (1998) 'Capital, state and space: contesting the borderless world', *Transactions of the Institute of British Geographers*, 23 (3): 291–309.

six

Alternative Employment Spaces

Colin C. Williams and Jan Windebank

As outlined in the Introduction to this book, the notion of an 'alternative economic space' is a slippery concept. How one defines an 'alternative' economic space often depends upon what one perceives as constituting the mainstream. At the outset, therefore, let us be clear that we are defining the mainstream here as 'formal employment', by which we mean paid work that is recorded in the official statistics. For us, this is the 'mainstream' because work that takes place under the social relations of formal employment is not only the subject matter of most economic geographical enquiry but also the chief focus of economic and spatial policy.

Given this definition of the mainstream, we here view 'alternative economic spaces' to comprise all those heterogeneous forms of work that lie beyond employment. Often referred to as the 'informal' sector/economy/sphere, to differentiate it from the mainstream 'formal' economy/sector/sphere, this informal domain comprises economic activities that involve the non-market production, consumption or exchange of goods and services. In other words, the 'alternative' sphere is being defined by what it is not; it is the 'other' work that is not formal employment. Adopting a dualistic 'either/or' mode of thinking to discuss the 'other' (that is, informal activity) that is not formal employment is problematic. It fails to recognize the diversity of practices on both sides of the equation. For example, by lumping everything that is not formal employment into a catch-all 'informal' sphere, the fact that there are many diverse forms of activity that exist beyond formal employment is seen as unimportant. However, no other definition of the 'informal' is possible. It is work beyond employment. As such, this definition will have to suffice.

Nevertheless, what we can do here is to unpack the 'informal' in order to recognize the heterogeneity within this sphere. To do this, we follow convention by dividing it into three slightly more coherent categories (for example, Leonard, 1998; Pahl, 1984), although there is still great diversity within each type of work. First, there is 'paid informal work'. This covers all of the paid production and sale of goods and services that are unregistered by or hidden from the state for tax, social security and/or labour law purposes but which are legal in all other respects (Commission for the European Communities, 1998; Portes, 1994; Thomas, 1992; Williams and Windebank, 1998). Secondly, there is 'self-provisioning', which is the unpaid household work undertaken by household members for themselves or for other members of their household. Finally, there is 'mutual aid', which is unpaid work done by household members for members of households other than their own. Although each of these categorizations is, as will be illustrated below, still perhaps being asked to do too much in that each covers a diverse array of social relations, it is this standard classification of the informal sphere that will be used here.

With this definition of the informal sphere in hand, the aim of this chapter is to start to map the uneven economic geographies of informal work. This remains a relatively uncharted territory. Despite the hard-fought battle of feminist geographers to get informal work recognized and valued (for example, England, 1996; Hanson and Pratt, 1995; Katz and Monk, 1993; McDowell, 1991, 1993; Women and Geography Study Group, 1997), the study of the informal economy is often viewed as little more than an addendum to the mainstream study of the formal sphere in economic geographical enquiry (see, for example, Bryson et al., 1999; Clark et al., 2000; Lee and Wills, 1997). Contesting this bias towards employment in economic geography is not helped by those who justify their study of the informal sphere in terms of how they are enhancing knowledge of the formal sphere. Understanding gender divisions of domestic work, for example, is sometimes asserted to be an essential prerequisite for understanding socio-spatial disparities in employment (see Hanson and Pratt, 1995). Here, however, we argue that the study of the informal sphere, similar to formal employment, is important in itself. No studies of formal employment justify their enquiry in terms of how it is a prerequisite for understanding the informal sphere. Why, therefore, should studies of the informal sphere justify themselves in terms of their importance for understanding formal economic geographies? Part of the project of studying alternative economic spaces is to decentre the mainstream from its core position. We thus need to stop justifying our investigation of alternative economic spaces in terms of how they provide a fuller understanding of the centre, in this case 'formal employment'.

TABLE 6.1 The trajectory of economic development: unpaid work as a percentage of total work time, 1965, 1975, 1985 and 1995

	1965	1975	1985	1995
UK[a]			48.1	58.2
France[b]		52.0	55.0	57.5
USA[c]	56.9	57.6	58.4	

a. Gershuny and Jones (1987); Murgatroyd and Neuburger (1997).
b. Chadeau and Fouquet (1981); Dumontier and Pan Ke Shon (1999); Roy (1991).
c. Robinson and Godbey (1997).

Indeed, for us, a crucial question is whether employment can and should any longer be seen as the 'centre'. Studies of the amount of time that people spend engaged in employment and unpaid work, for example, raise doubts about the logic of placing employment at the centre and informal work at the margins (see Table 6.1). Although the total amount of time spent engaged in work is declining in the United Kingdom, France and the USA, the total time spent in employment is declining faster than the time spent in unpaid work. In consequence, rather than a formalization of the advanced economies, as is often assumed, the past three decades have witnessed a process of informalization, at least in terms of the time spent engaged in unpaid work relative to paid work. The most recent data, moreover, reveal that the total time spent engaged in unpaid work is now greater than the total time spent engaged in employment. To continue to focus upon employment therefore is arguably to concentrate on studying a residual and diminishing sphere of economic activity. The 'mainstream' it seems, may no longer be quite what we previously assumed.

Given that the informal sphere constitutes the place where the majority of time spent working by the populations of the advanced economies is passed, it seems to us absolutely essential that greater effort is put into understanding *the uneven economic geographies of work beyond employment*. Presently, and in contrast to the wealth of knowledge on how formal employment varies across space, there is little understanding of how the extent and nature of informal work varies across space or how the motives underlying such work vary spatially. Are there spatial variations in the extent to which informal work is used? In other words, is it more extensively employed in some areas than others? And is the way in which populations participate in informal work the same everywhere? Or is it the case that in some areas the nature of such work varies? And finally, are the motives for engaging in such work universal? Or is it the case that it is conducted more out of preference in some areas and necessity in others?

To answer these questions, this chapter reports the results of interviews with 511 households in a variety of higher- and lower-income urban neighbourhoods in two contrasting English cities during 1998–99. First, we map these uneven economic geographies in terms of the unequal capability of households to perform necessary work along with the overall household work practices that they use. With this in hand, and starting with self-provisioning, followed by mutual aid and paid informal work, we then evaluate the spatial variations in the extent of participation in each form of work, its nature and the motives of people participating in this realm of work. This will reveal the complex ways in which participation in informal work varies across space.

EXAMINING SPATIAL VARIATIONS IN THE PERFORMANCE CAPABILITIES AND PRACTICES OF HOUSEHOLDS

Given that previous research highlights a strong relationship between income levels and participation in informal economic activities (for example, Fortin et al., 1996; Pahl, 1984; Renooy, 1990; Van Geuns et al., 1987; Williams and Windebank, 1998), higher- and lower-income cities and neighbourhoods were chosen for this study of informal work. First, therefore, two contrasting cities were selected. Southampton is a relatively affluent 'southern' city that has witnessed strong service growth, resulting in low unemployment rates and high average wage rates. By contrast, Sheffield is a relatively deprived 'northern' city witnessing de-industrialization and only weak service sector growth, resulting in high unemployment and low wage rates. Green and Owen (1998) identified both inner-city neighbourhoods with relatively high concentrations of ethnic minorities and 'sink' council estates as the locality-types suffering the largest increases in economic inactivity, non-employment and unemployment. Consequently, an example of these two types of neighbourhood was chosen in each city as representative of lower-income neighbourhoods. In addition, an affluent suburb was selected in the two cities. In each lower-income neighbourhood, 100 households were interviewed, resulting in 400 interviews in total. In the two higher-income neighbourhoods, meanwhile, 50 and 61 households were interviewed in the southern and northern city respectively (resulting in a total of 511 interviews).

Structured interviews based on a list of 44 common household services[1] generated from Pahl's (1984) study of the Isle of Sheppey were used to explore the ability of households to perform necessary work and the work practices that they used. In contrast to Pahl's study, however, these interviews also sought to identify whether a task was seen as necessary, the motivations of suppliers and consumers, along with employment status, gender, age and relationship of the customer to the supplier. Previous research using this technique reveals that when the results from households as customers and suppliers are compared, the same levels of unpaid and paid informal exchange are identified, meaning that the technique does not suffer from under- or over-reporting by respondents on the supply or demand side (for example, Leonard, 1994; Pahl, 1984). Indeed, this was also found in the survey reported here, suggesting that the data are relatively reliable.

Here, therefore, an evaluation is undertaken of first, the coping abilities of households in these different cities and neighbourhoods, secondly, how household work practices vary across space, and thirdly, the spatial variations in self-provisioning, mutual aid and paid informal exchange.

SPATIAL VARIATIONS IN THE PERFORMANCE CAPABILITIES OF HOUSEHOLDS

Much previous research on spatial inequalities has focused upon how the inputs into households (for example, gross household income) vary across space. Here, however, we adopt an 'outputs' approach by examining the capabilities of households to perform work that they perceive as necessary. In so doing, we follow the approach advocated by Amartya Sen (1995, 1999), who argues that inequality needs to be analysed not in terms of opulence (for example, 'real income' estimates) or utility (as in welfare economic formulations), but in terms of the 'capabilities' of a person or household. For our purposes here, this approach is useful because it recognizes the contribution of all forms of work to coping practices. Some households and areas, for example, although money-poor, might be able to draw upon deep and wide social support networks in order to perform necessary work. As such, their outputs will be far greater than other households with a similar income level but lacking such support networks.

The first column of Table 6.2 denotes the contrasting capabilities of households in different areas to get the 44 tasks completed. This reveals that the ability of households to perform necessary tasks is not uniform across space. There is a 'wealth gap' between higher- and lower-income neighbourhoods. This is not surprising. The 'wealth gap' between higher- and lower-income neighbourhoods is wider in Southampton than Sheffield (Williams, 2001). In Southampton, households in the affluent suburb completed 57.3 per cent of all the tasks surveyed compared with 45.3 per cent in the lower-income neighbourhoods. In Sheffield, by contrast, although households in the affluent suburb again completed a higher proportion of tasks (53.3 per cent), the gap with households in the lower-income neighbourhoods (who completed 49 per cent of tasks) was not so great as in Southampton.

However, this does not take into account whether households deemed these uncompleted tasks to be in need of performing. In Southampton, households in the affluent suburb wanted to undertake just 24 per cent of the uncompleted tasks, but in the lower-income neighbourhoods of this city, this figure was 66 per cent. In Sheffield, in contrast, this gap in necessity is narrower: in the affluent suburb, households wished to undertake 32 per cent of the uncompleted tasks while this figure was 60 per cent in the lower-income neighbourhoods. Thus, the gap between the higher- and lower-income neighbourhoods is wider in Southampton than Sheffield in terms of their coping capabilities. However, Southampton lower-income neighbourhoods are less able to perform necessary tasks than their Sheffield counterparts.

Why is this the case? Here, the suggestion is that these findings can be tentatively explained by exploring the interrelationships between wage rates, cost of living and social payments (for further analysis, see Williams, 2001). On the one hand, the cost of living in Sheffield is 9.5 per cent less than in Southampton, measured in terms of the cost of a basket of goods and services (Reward Group, 1999). Despite this, households reliant on social payments (for example, pensions and welfare benefits) receive the same wherever they live. The result is that Southampton jobless households are worse off than their Sheffield counterparts in terms of their ability to get necessary tasks completed (see Williams, 2001). On the other hand, average wage rates in Southampton are not only higher than in Sheffield, but the variation is greater than the difference in the cost of living (Reward Group, 1999). The outcome is a wider 'wealth gap' in Southampton than Sheffield between higher-income neighbourhoods, where most are in employment, and lower-income neighbourhoods, where many are dependent on state social payments.

Geographical variations in the cost of living and/or *real* income levels and national-level payments (for example, the national minimum

TABLE 6.2 Extent and nature of household work practices (by geographical area)

	Tasks conducted (%)	Self-provisioning		Unpaid mutual aid		Paid informal exchange		Formal employment	
		Mean	(%)	Mean	(%)	Mean	(%)	Mean	(%)
Southampton									
Lower-income inner-city neighbourhood	43.1	14.2	75.0	0.6	3.4	0.8	4.2	3.3	17.4
Lower-income council estate	47.4	15.5	74.5	0.8	3.8	0.9	4.5	3.6	17.2
Both lower-income areas	45.3	14.9	74.8	0.7	3.6	0.9	4.4	3.4	17.3
Higher-income suburb	57.3	18.0	71.3	0.5	1.9	1.6	6.5	5.1	20.3
Sheffield									
Lower-income inner city neighbourhood	49.7	16.6	75.6	0.9	4.2	1.4	6.4	3.0	13.8
Lower-income council estate	48.3	16.8	79.3	0.8	3.7	0.9	4.3	2.7	12.7
Both lower-income areas	49.0	16.7	77.4	0.9	3.9	1.2	5.4	2.9	13.3
Higher-income suburb	53.3	17.1	72.8	0.4	1.9	2.6	11.2	3.3	14.1

wage, tax credits, social payments) have significant socially-reproductive effects and raise the question of whether such payments might be locally and/or regionally differentiated. However, such issues have so far received little attention by economic geographers (for example, Sunley and Martin, 2000; Williams, 2001). It is not only these economic processes, however, that require further attention if the geographies of work are to be more fully understood. Although economic explanations may tentatively be used to explain the spatial variations in the coping capabilities of households, such economic essentialism is not always sufficient to explain the spatial variations in the extent and nature of self-provisioning, mutual aid and paid informal work.

SPATIAL VARIATIONS IN HOUSEHOLD WORK PRACTICES

Examination of the variations in the extent to which each form of work is used in different areas reveals some surprising findings. Higher-income neighbourhoods tend to be more formalized and money-oriented in their work practices than lower-income areas. This is perhaps to be expected. Households in higher-income areas are more able to pay others to do work for them than households in lower-income areas. However, although households in more affluent areas are more heavily reliant on formal sources of labour and monetized exchange to get work completed, the number of tasks that they conducted through self-provisioning was actually greater than in lower-income neighbourhoods. This finding will be examined further below. For the moment, it is simply important to note that affluent suburbs conduct greater levels of self-provisioning than lower-income neighbourhoods.

Even more surprising, however, were the differences in household work practices between the two cities. Superficially, the findings do not appear controversial. Households in the affluent southern city relied more heavily upon formal sources of labour to get work done compared with their northern counterparts. This is similar to the finding between affluent and deprived neighbourhoods described above. But what is different is that households in the more affluent southern city do not also engage in a wider range of self-provisioning. And yet previous studies of social groups have shown that higher-income households conduct greater amounts of informal work than lower-income households (for example, Leonard, 1998; Pahl, 1984; Renooy, 1990; Williams and Windebank, 1999) and our findings show that this is also true of affluent and deprived neighbourhoods. To explain this scalar discrepancy between rich and poor cities and rich and poor neighbourhoods, we here turn our attention to unpacking household work practices. First, we analyse the spatial variations in self-provisioning, followed by mutual aid and paid informal work. In each case, we explore the variations between affluent and deprived neighbourhoods as well as between the two cities in terms of both the levels and character of each form of informal work and the motives underlying participation in such work. This will reveal the complex ways in which participation in informal work varies across space.

SPATIAL VARIATIONS IN THE PRACTICE OF SELF-PROVISIONING

To explain the propensity of affluent suburbs to engage in a slightly greater range of tasks using self-provisioning than lower-income neighbourhoods, which appears to contradict the popular perception that the affluent buy-in formal labour in order to relieve them of the need to engage in self-provisioning, two factors are important. On the one hand, household type is a key factor. In lower-income neighbourhoods, households are more likely to be composed of the sick/disabled or the retired, to be in rented accommodation and to be lower skilled. They are thus less likely to possess the necessary combination of money, tools, confidence, knowledge, practical skills and physical ability, and/or to have the responsibility to undertake self-provisioning. To take just one example, the ability to paint one's outside windows requires money (to buy the paint and brushes), tools (for example, a ladder), the confidence, knowledge and practical skills to perform the task, a certain physical ability (for example, to climb ladders) and, finally, responsibility for doing so. If any of these are lacking, which is often the case in lower-income neighbourhoods, the task cannot be carried out using self-provisioning. In affluent suburbs, meanwhile, such assets are seldom lacking.

On the other hand, the greater propensity to engage in self-provisioning in affluent suburbs also has to be understood in the context of the types of task performed and the motives of the participants. Households in lower-income neighbourhoods are more likely to conduct those tasks perceived as essential and/or routine work tasks, while for households in higher-income areas, a greater proportion is voluntarily chosen and conducted out of choice rather than necessity. Different places, therefore, have contrasting preference/necessity ratios when conducting self-provisioning. This shapes, and is shaped by, attitudes towards self-provisioning and also how such work is interpreted in practical day-to-day contexts in different neighbourhoods. Although 44 per cent of the tasks undertaken using self-provisioning in the lower-income neighbourhoods was primarily motivated by economic necessity, this was the case for just 10 per cent in the higher-income neighbourhoods. The reason self-provisioning was conducted out of preference was either due to ease (18 per cent and 36.9 per cent of all self-provisioning in lower- and higher-income neighbourhoods respectively), desire to individualize the end-product (21 per cent and 24 per cent of all such work respectively) or the pleasure got from doing the work (14 per cent and 32 per cent respectively).

Hence, despite the greater penetration of formal sources of labour in affluent suburbs (to get some of the routine work conducted), this does not result in these populations doing less self-provisioning. Instead, the greater ability to buy-in services from the formal sphere (for example, cleaning, childcare) appears to release them to conduct a wider range of self-provisioning. These tasks then tend to be conducted out of preference rather than necessity. The employment of a cleaner or nanny, for instance, seems to release these households (particularly women) to participate in self-provisioning in a different, more voluntary way than in lower-income neighbourhoods. Put another way, and reflected in the growth of DIY television programmes, there is a *Changing Rooms* effect,[2] especially in affluent populations, whereby households engage in DIY jobs voluntarily for the pleasure they get from doing the task. The result is that participation in self-provisioning has a contrasting logic in different neighbourhoods. There is a greater degree of human agency prevalent in affluent suburbs while economic necessity is the dominant rationale in lower-income neighbourhoods.

Comparing the two cities meanwhile, a second issue that needs to be explained is the slightly lower propensity to use self-provisioning in Southampton and the much lower desire to engage in self-provisioning in this southern city compared with its northern counterpart. The lower propensity to engage in self-provisioning in this southern city can be explained by two factors. On the one hand, the lower-income neighbourhoods in Southampton are worse off, measured in terms of their ability to perform necessary work, and are thus unable to do the same level of self-provisioning as their Sheffield counterparts (that is, 14.9 tasks in Southampton compared with 16.7 in Sheffield, or 12 per cent fewer tasks). On the other hand, the Southampton affluent suburbs are better off, again measured in terms of their performativity, and thus formalize a greater range of activity (20.3 per cent of all tasks compared with 14.1 per cent).

However, these factors do not explain why Southampton households expressed a greater desire to formalize their self-provisioning work than their Sheffield counterparts. Some 72 per cent of Southampton households asserted that they would do less self-provisioning if they had more money compared with just 32 per cent in Sheffield. For us, this is an indicator of contrasting regional 'work cultures'. The work practices displayed in Table 6.2 showing a slightly greater tendency towards formalized and/or monetized production in Southampton than Sheffield are thus just one minor reflection of a much wider rift in work cultures between the two cities. It signifies a so far unrecognized but fascinating difference in *attitudes* towards self-provisioning. This geographically differentiated propinquity towards formalization as well as self-provisioning will require far more research, however, before it can be stated with any certainty. Whether there are differences between southern and northern cities in general, for example, and between urban areas and the countryside where there is perceived to be more of a 'self-reliance ethic' (for example, Shucksmith, 2000), requires much greater investigation.

SPATIAL VARIATIONS IN THE PRACTICE OF MUTUAL AID

Is it only the extent, character and motives for self-provisioning that varies across space or is it the same when mutual aid is examined? Does the practice of mutual aid differ geographically? And does the logic underlying its usage also vary? This work, exchanged on an unpaid basis within the extended family and social or neighbourhood networks, occupies the ground between self-provisioning and more formal institutions such as private, public and 'third sector' organizations charged with delivering goods and services either on a paid, statutory or volunteer basis. In other words, it is what Bulmer (1989: 253) refers to as an 'intermediary structure' that is beyond the realm of household relations but more familiar to us than the impersonal institutions in the wider society. Mutual aid can thus be seen as a constituent component of 'social capital', defined as those 'features of social organisation such as networks, norms and social trust that facilitate co-ordination and co-operation for mutual benefit' (Putnam, 1993: 67). It is micro-level exchange that arises out of, and helps constitute, social networks and norms from which trust may derive (see Williams and Windebank, 2001a).

So far as the geographies of mutual aid are concerned, two mutually exclusive narratives pervade contemporary discourse. On the one hand, there remains a dominant perception that lower-income neighbourhoods are composed of solidaristic working-class populations who have a greater tendency to help each other out (see, for example, Home Office, 1999). This perception dates back to work conducted over a quarter of a century ago that highlighted the degree of mutual aid in such neighbourhoods (for example, Young and Wilmott, 1975). On the other hand, there is a newer narrative of 'sink' estates, composed of perceptions of acute crime and fear of crime, the loss of community spirit and a sense of decline so far as quality of life is concerned, as populations become locked into these increasingly debilitated and alienated communities (for example, Social Exclusion Unit, 1998).

The overwhelming finding of our study is that the reality in lower-income areas is closer to the second narrative than the first. Although a greater proportion of work is performed using mutual aid in lower-income than in affluent areas (see Table 6.2), this does not signify the continuing presence of a solidaristic working-class community in such places. As will be illustrated below, much of this exchange is undertaken out of necessity rather than choice.

In higher-income neighbourhoods, just 6.8 per cent of exchanges take place on an unpaid basis compared with 15.6 per cent in lower-income neighbourhoods. This greater propensity to engage in exchange on an unpaid basis in lower-income communities, however, has little to do with the persistence of an ethic of mutual aid. Analysing the source of such mutual aid, the finding in both the affluent and deprived neighbourhoods is that it is mostly conducted for and by kin. In the deprived urban neighbourhoods, some 70.3 per cent of mutual aid is provided by relatives, 26.6 per cent by friends/neighbours and just 3.1 per cent by voluntary organizations. In affluent suburbs, meanwhile, there is a greater reliance on friends/neighbours (44 per cent) and relatively less on relatives (56 per cent). This difference in the constitution of mutual aid, however, is almost entirely due to households in these areas being more likely to have kin in the city upon whom they call for help and give aid (see Table 6.3).

Hence, mutual aid may be more prevalent in deprived urban neighbourhoods, but this is due to households having more kin in the city. Indeed, outside kinship exchange, there is not only relatively little mutual aid taking place, but also little desire to engage in such aid. Although most interviewees were willing to engage in mutual aid for kin, they overwhelmingly asserted that others outside the family network could look after themselves so far as they were concerned. Providing help for the family, however, was different. Unpaid kinship exchange was widely supported in these lower-income populations,

TABLE 6.3 Percentage of households with kin living in the city (by neighbourhood type)

	Lower-income neighbourhoods	Higher-income neighbourhoods
Grandparents	20.4	5.4
Parents	67.2	22.5
Brothers or sisters	67.2	31.6
Children	68.4	51.4
Uncles or aunts	54.5	15.3
Cousins	60.6	15.3

frequently referred to as 'done out of the kindness of our heart', 'we like to help out', 'there were family reasons' or 'we did it out of love'. Indeed, the only factor constraining both the giving and receiving of kinship support was the fact that many households, as shown above, had few kin living in the city.

Beyond kinship exchange, however, and unlike earlier studies (for example, Young and Wilmott, 1975), there was found to be little desire to help others on an unpaid basis. Indeed, there were only four circumstances where non-kinship unpaid exchange occurred: first, when it was felt to be unacceptable to pay somebody (for example, when they lent you something); secondly, when it was felt inappropriate to pay them (for example, when a colleague from work did you a favour); thirdly, when payment was impossible (for example, when somebody refused to be paid because they wanted an unpaid favour from you at a later date), and finally, when the social relations mitigated against payment (for example, when the recipient could not afford to pay and thus had no choice but to offer a favour in return). If at all feasible, however, interviewees avoided unpaid exchange. Paying people was by far the preferred option. Monetary exchange avoids any obligation 'hanging over you' to reciprocate favours but, at the same time, the wheels had been oiled for the maintenance or creation of closer relations through exchange without being 'duty bound'. Seen in this light, the only reason for the greater prevalence of unpaid exchange in lower-income neighbourhoods is due to the higher level of kin in the vicinity, leading to greater levels of unpaid kinship exchange and the inability of households in lower-income neighbourhoods to pay for help rendered. Unpaid exchange was a necessity, not some choice borne out of the existence of trust and/or close social relations (see Putnam, 1995).

Another popular prejudice in the UK so far as the geographies of mutual aid are concerned is that northerners help each other out to a greater extent than southerners (that is, there is a stereotype of the helpful, friendly northerner and cold, detached southerner). This study finds that the northern populations studied in Sheffield are slightly more likely to use mutual aid (see Table 6.2). One might very tentatively conclude that this view thus has some grounding in reality. However, this should not hide the fact that in both cities there is a marked reluctance for people to go out of their way to offer material support to non-kinship ties, especially in lower-income neighbourhoods.

In sum, the old-fashioned idea of tight, self-sufficient communities helping each other out appears to be only marginally relevant today (see Burns and Taylor, 1998). People do not want to use mutual aid and do so only when they cannot do the job themselves, afford to pay somebody, or when the social relations involved militate against payment. Consequently, it is for these reasons and these reasons alone that some areas have a greater propinquity towards unpaid mutual aid.

SPATIAL VARIATIONS IN THE PRACTICE OF PAID INFORMAL WORK

What, however, of paid informal work? Is it the case, as popular prejudice asserts, that this form of work is more prevalent in poorer localities and regions? To examine this, it is first necessary to outline the social bases of this viewpoint. Such work is seen to be more prevalent in lower-income neighbourhoods because it has been viewed as an activity that is performed for purely economic motives. In other words, paid informal work is seen to exemplify the more flexible, profit-motivated exchange relations that have arisen under post-Fordist regimes of accumulation (Castells and Portes, 1989; Leonard, 1994; Portes, 1994; Sassen, 1989). Conceptualized as a form of low-paid peripheral employment conducted by marginalized groups for unadulterated economic reasons (for example, Castells and Portes, 1989; Kesteloot and Meert, 1999; Portes, 1994), this 'marginality thesis' assigns such work to marginalized areas. Here, however, we will show that such an economic reading of paid informal work is a misrepresentation of its meaning and, thus, that its geographical distribution is different from what has often been presupposed (for an in-depth examination, see Williams and Windebank, 2001b).

As with previous studies (for example, Fortin et al., 1996; Leonard, 1994; Pahl, 1984; Renooy, 1990), this study of English cities confirms the concentration of paid informal work in higher-income populations (see Table 6.1). Indeed, it is not just customers of paid informal work that are concentrated in affluent suburbs. The populations of lower-income neighbourhoods (78.2 per cent of the sample) supply just 62.5 per cent of the paid informal work, while residents of affluent suburbs (21.8 per cent of the sample) supply 37.5 per cent of such work. Comparing lower- and higher-income areas, the average amount received for conducting a task on a paid informal basis was £90.24 compared with £1,665 respectively; the average hourly informal wage rate was £3.40 compared with £7.50; and the mean annual household income from paid informal work was £46.22 compared with £435.62 (see Williams and Windebank, 2001b).

However, our findings do not concur with the current perception that such work is everywhere economically motivated. Instead, it finds that although such work is firmly embedded in unadulterated economic motives in affluent suburbs, this is not the case in lower-income neighbourhoods. In these areas, much paid informal exchange is conducted for and by close social relations for primarily social reasons or to help each other out in a way that avoids any connotation of charity and it is conducted more by women than men. In higher-income areas, meanwhile, such exchange is conducted more by self-employed people and firms for profit, is primarily used as a cheaper alternative to formal firms and is undertaken more by men than women (for an in-depth examination, see Williams and Windebank, 2001c).

In lower-income neighbourhoods, less than a third (31.7 per cent) of all paid informal exchanges used firms or unknown self-employed individuals, but such sources constituted 84 per cent in higher-income areas. Therefore, in lower-income neighbourhoods, paid informal exchanges mostly involve transactions conducted by friends, neighbours or relatives, while it is between anonymous buyers and sellers in higher-income neighbourhoods. This has a major impact on the motivations of purchasers. In some 80 per cent of circumstances where paid informal work was used in affluent suburbs, it was employed as a cheaper alternative to formal employment, but this is the reason for just 18.1 per cent of this work in lower-income neighbourhoods. In nearly all of these cases, it is firms and/or self-employed people not known by the household who conducted the work. When closer social relations are involved, it is either carried out for social reasons or seen as an opportunity to give money to another person in a way that avoids any connotation of charity (see Kempson, 1996).

Social reasons tend to predominate when friends or neighbours (rather than relatives) are involved. Indeed, it is in these paid informal exchanges between friends and neighbours that one can see the much-discussed mutual aid that supposedly used to take place on an unpaid basis. For participants, these exchanges are conducted in order to cement or consolidate social relationships. The exchange of cash is seen as a necessary medium, especially when neighbours or friends are involved, because it prevents such relations from turning sour if and when somebody reneges on their commitments. Cash thus provides the oil to allow mutual aid in situations where trust is missing. Indeed, many of the respondents, especially in the deprived neighbourhoods, could not remember such relationships ever being any different. This, therefore, is not evidence of the formalized monetization of previously unpaid reciprocal exchange. It is evidence that cash exchange does not have to be profit-making in orientation and can be grounded in alternative social relations. As Hart (2000) puts it, market relations do not necessarily have to be capitalist. And it is here in these deprived neighbourhoods that solid evidence is available of the existence of monetized, market-based relations devoid of profit-motivated intentions and capitalist social relations.[3]

Besides social reasons, another rationale for paid informal exchange is redistribution. This mostly occurs when kin are involved. In these instances, it was either children or a relation such as a brother, sister or parent who was paid, normally in order to give them much needed spending money, when for example, they were unemployed. Indeed, using relatives to do tasks so as to give them money was the principal rationale behind 10.9 per cent of all paid informal work in the lower-income neighbourhoods. For these consumers, therefore, it was a way of giving money to a poorer relative in a way that avoided all connotations of charity, even if this was an intention underlying such exchange. This type of motivation, however, hardly existed in affluent suburbs, except when young children were involved who were paid by their parents to do a job.

It is not only between higher- and lower-income neighbourhoods, however, that the ways in which people participate in paid informal work, and their reasons for doing so, vary. It is also the case when Sheffield and Southampton are compared. In Sheffield, a much higher proportion of paid informal work is conducted through relatives, friends and neighbours than in Southampton. For example, in the higher-income Southampton suburb, 92.1 per cent of paid informal exchange was conducted using people previously unknown to the customer, but only 76.7 per cent in the Sheffield suburb. The result is that this work was more likely to be conducted for profit-motivated reasons in Southampton than Sheffield. In part, this is due to the relative inability of Southampton households to pay formal labour and in part due to their greater preference to engage in monetary relations rather than either do-it-themselves or seek unpaid help.

In sum, paid informal exchanges are largely driven by non-market motivations in lower-income neighbourhoods (and Sheffield) but by the profit motive in affluent suburbs (and Southampton). As with self-provisioning, therefore, participation in paid informal work appears to have a varying logic in different types of neighbourhood and city. Indeed, these contrasting local logics mean that, although the profit motive has deeply penetrated monetary relations in affluent suburbs, some 30 per cent of all the monetary exchanges (both formal and informal) studied in these lower-income neighbourhoods were conducted for rationales beyond the profit motive.

The implications of this finding are potentially profound. It shows that despite the penetration of monetary relations into every nook and cranny of social life (see Harvey, 1989; Leyshon, 1997; Sayer, 1997), these are by no means everywhere driven by the social relations of capitalism. Monetary exchange is not everywhere based on profit-motivated market relations. Indeed, this study of paid informal exchange, often taken as an exemplar of profit-motivated monetary exchange, displays how it is wholly feasible to have monetary transactions that are embedded in alternative social relations and motives. This raises the fascinating questions of whether similar social relations and motives are found more widely in the formal economy and whether some places are more embedded in such social relations and motives than others. Although there are now pointers that at least some formal spheres of activity may well indeed be operating under non profit-motivated social relations and motives (for example, Lee, 2000), the investigation of these issues remains in its infancy.

CONCLUSIONS

How does the foregoing analysis further our understanding of alternative economies? This chapter has taken formal employment as the mainstream and explored the alternative economic space of the informal sphere. In so doing, it has shown that, although most scholars assume that employment is the principal form of work in contemporary society, this is not the case, at least so far as the time spent engaged in formal and informal work is concerned. A crucial finding of this chapter, therefore, is that employment can and should no longer simply be accepted without question as being at the 'centre'. Until such time as those economic geographers whose work is biased towards the study of employment justify why this should be the case, then doubts must remain about whether it is any longer acceptable to focus upon employment without justification. In terms of time spent engaged in such work, it is a 'marginal' form of work.

A further contribution of this chapter to the alternative economic spaces literature has been to show the uneven economic geographies of the informal sphere in terms of its extent, character and motives. Although the spatial variations in the magnitude of the informal sphere are not always marked, the character of these alternative economies and the logics underlying their use display major spatial variations. Take, for example, self-provisioning. In affluent suburbs, although participation in self-provisioning is slightly greater in higher- than in lower-income neighbourhoods, the work conducted is of a very different nature in that a higher proportion is freely chosen activity conducted for pleasure. In lower-income neighbourhoods, in contrast, self-provisioning is composed of more routine tasks conducted for reasons of economic necessity rather than out of preference. Similarly, there are stark geographical contrasts in the nature of, and logic for, paid informal work. In affluent suburbs, paid informal work conforms to the conventional view of such exchange in that it is conducted for the purpose of profit, mostly for people who are barely known to the supplier. In lower-income neighbourhoods, however, such work is largely undertaken for relatives, friends and neighbours for the purposes of sociability and redistribution. In other words, it is monetized mutual aid. Indeed, and to synthesize our findings, a tentative conclusion would be that the affluent suburbs studied seem to be more formalized, money-based and profit-motivated in their work practices. Concomitantly, the economies of the lower-income neighbourhoods studied appear to be less formalized, less money-based and far less profit-motivated.

Similarly, when the two cities are compared, the Southampton 'economy' appears more formalized money-oriented and the profit motive more prominent while the Sheffield 'economy' appears relatively informalized, less money-oriented and profit-seeking. For some, this difference in household work practices in Southampton and Sheffield might be explained simply in terms of economic forces. The relatively deprived city of Sheffield and affluent city of Southampton portray the same characteristics as deprived and affluent neighbourhoods. However, before reaching such a conclusion, it is important to note that Southampton households display less desire to engage in self-provisioning for pleasure. They also show a greater preference for engaging in monetary exchanges rather than either doing-it-themselves or seeking unpaid help. The only conclusion that can be reached, therefore, is that there may well be distinct 'cultures of work' in different places. Examining informal work therefore leads to new possibilities for understanding contemporary economic geographies. It may well be that there are distinct differentiated economic cultures prevailing in contemporary Britain that have been completely missed by limiting analysis to the geographies of formal employment.

Some may assert that this is all very interesting but has few implications for policy. Indeed, this retort is increasingly raising its head in economic geographical discourses (for example, Martin, 2000). Hopefully, however, we have displayed that there are major policy implications arising out of the study of these alternative economic spaces. Examining household coping capabilities, this chapter has shown how populations living in different areas have contrasting abilities to perform necessary everyday tasks. We have revealed first, a wider gap between the higher- and lower-income neighbourhoods in Southampton than Sheffield, and secondly, how Southampton lower-income neighbourhoods are less able to perform necessary work than their Sheffield counterparts. These findings have been explained in terms of the interrelationships between wage rates, cost of living and social payments. Despite the cost of living being higher in Southampton than that in Sheffield, real wage rates more than compensate for this difference. Those in employment are thus better off in Southampton, in terms of their ability to perform necessary work, while those reliant on nationally determined social payments (for example, pensions, welfare benefits) are worse off. The result is a wider 'wealth gap' in Southampton than Sheffield between higher-income neighbourhoods (where most are in employment) and lower-income neighbourhoods (where many are dependent on state social payments). Economic geography thus needs to investigate further the relationships between geographical variations in the cost of living and *real* incomes, as well as whether national-level payments (for example, the national minimum wage, tax credits, social payments) might be locally and/or regionally differentiated.

This study of alternative economic spaces thus provides both new ways of reading uneven economic geographies and a number of policy implications that require further consideration. Just as importantly, it shows that alternative economic spaces are not everywhere the same. The informal sphere is constituted in very different ways, and used for varying purposes, in different places. It seems likely that the same will apply to all alternative economic spaces, however defined.

Notes

This research was funded by the European Commission's DG12 under its Targeted Socio-Economic Research programme and the Joseph Rowntree Foundation, which supported this project as part of its programme of research and innovative development projects, which it hopes will be of value to policy-makers and practitioners. However, the facts presented and views expressed in this chapter are those of the authors. The authors would like to thank Stephen Hughes and Jo Cook for providing the research assistance.

1 The 44 tasks examined covered house maintenance (outdoor painting; indoor painting; wallpapering; plastering; mending a broken widow and maintenance of appliances), home improvement (putting in double glazing; plumbing; electrical work; house insulation; putting in a bathroom suite; building a garage; building an extension; putting in central heating and carpentry), housework (routine housework; cleaning windows outdoors; spring cleaning; cleaning windows indoors; doing the shopping; washing clothes and sheets; ironing; cooking meals; washing dishes; hairdressing; household administration), making and repairing goods (making clothes; repairing clothes; knitting; making or repairing furniture; making or repairing garden equipment; making curtains), car maintenance (washing car; repairing car and car maintenance), gardening (care of indoor plants; outdoor borders; outdoor vegetables; lawn mowing) and caring activities (day-time baby-sitting; night-time baby-sitting; educational activities; pet care).
2 *Changing Rooms* is a popular UK television programme in which friends and neighbours redecorate a room in each other's house under the direction of a professional interior designer.
3 For details of how the pricing mechanisms for such paid informal work operate, see Williams and Windebank (2001b).

References

Bryson, J., Henry, N., Keeble, D. and Martin, R. (eds) (1999) *The Economic Geography Reader: Producing and Consuming Global Capitalism*. Chichester: Wiley.

Bulmer, M. (1989) 'The underclass, empowerment and public policy', in M. Bulmer, J. Lewis and D. Piachaud (eds), *The Goals of Social Policy*, London: Unwin Hyman. pp. 245–57.

Burns, D. and Taylor, M. (1998) *Mutual Aid and Self-Help: Coping Strategies for Excluded Communities*. Bristol: The Policy Press.

Castells, M. and Portes, A. (1989) 'World underneath: the origins, dynamics and affects of the informal economy', in A. Portes, M. Castells and L.A. Benton (eds), *The Informal Economy: Studies in Advanced and Less Developing Countries*. Baltimore, MD: Johns Hopkins University Press.

Chadeau, A. and Fouquet, A. (1981) 'Peut-on mesurer le travail domestique?', *Economie et Statistique*, 136: 29–42.

Clark, G.L., Gertler, M. and Feldman, M. (eds) (2000) *Handbook of Economic Geography*. Oxford: Oxford University Press.

Commission of the European Communities (1998) *On Undeclared Work*, COM (1998) 219. Brussels: Commission of the European Communities.

Dumontier, F. and Pan Ke Shon, J.-L. (1999) 'En 13 ans, moins de temps contraints et plus de loisirs', *INSEE Premiere 675*. Paris: INSEE.

England, K. (ed.) (1996) *Who Will Mind Baby? Geographies of Child Care and Working Mothers*. London: Routledge.

Fortin, B., Garneau, G., Lacroix, G., Lemieux, T. and Montmarquette, C. (1996) *L'Economie Souterraine au Quebec: Mythes et Realites*. Laval: Presses de l'Universite Laval.

Gershuny, J. and Jones, S. (1987) 'The changing work/leisure balance in Britain 1961–84', *Sociological Review Monograph*, 33: 9–50.

Green, A.E. and Owen, D. (1998) *Where are the Jobless? Changing Unemployment and Non-employment in Cities and Regions*. York: The Policy Press.

Hanson, S. and Pratt, G. (1995) *Gender, Work and Space*. London: Routledge.

Hart, K. (2000) *Money in an Unequal World*. London: Profile Books.

Harvey, D. (1989) *The Condition of Postmodernity: An Enquiry into the Origins of Cultural Change*. Oxford: Blackwell.

Home Office (1999) *Community Self-help*. London: Home Office.

Katz, C. and Monk, J. (1993) *Full Circles: Geographies of Women over the Life Course*. London: Routledge.

Kempson, E. (1996) *Life on a Low Income*. York: York Publishing Services.

Kesteloot, C. and Meert, H. (1999) 'Informal spaces: the geography of informal ecconomic activities in Brussels', *International Journal of Urban and Regional Research*, 23: 232–51.

Lee, R. (2000) 'Shelter from the storm? Geographies of regard in the worlds of horticultural consumption and production', *Geoforum*, 31: 137–57.

Lee, R. and Wills, J. (eds) (1997) *Geographies of Economies*. London: Arnold.

Leonard, M. (1994) *Informal Economic Activity in Belfast*. Aldershot: Avebury.

Leonard, M. (1998) *Invisible Work, Invisible Workers: The Informal Economy in Europe and the US*. Basingstoke: Macmillan.

Leyshon, A. (1997) 'Geographies of money and finance II', *Progress in Human Geography*, 21 (3): 381–92.

Martin, R.L. (2000) 'Geography and public policy: the case of the missing agenda', *Progress in Human Geography*, 25 (2): 189–210.

McDowell, L. (1991) 'Life without father and Ford: the new gender order of post-Fordism', *Transactions of the Institute of British Geographers*, 16: 400–19.

McDowell, L. (1993) 'Space, place and gender relations: feminist empiricism and the geography of social relations', *Progress in Human Geography*, 17: 157–79.

Murgatroyd, L. and Neuburger, H. (1997) 'A household satellite account for the UK', *Economic Trends*, 527: 63–71.

Pahl, R.E. (1984) *Divisions of Labour*. Oxford: Basil Blackwell.

Portes, A. (1994) 'The informal economy and its paradoxes', in N.J. Smelser and R. Swedberg (eds), *The Handbook of Economic Sociology*. Princeton, NJ: Princeton University Press.

Putnam, R. (1993) *Making Democracy Work: Civic Traditions in Modern Italy*. Princeton, NJ: University of Princeton Press.

Putnam, R. (1995) 'Tuning in, tuning out: the strange disappearance of social capital in America', *Political Science and Politics*, 28: 664–83.

Renooy, P. (1990) *The Informal Economy: Meaning, Measurement and Social Significance*. Netherlands Geographical Studies No. 115. Amsterdam: Netherlands Geographical Institute.

Reward Group (1999) *Cost of Living Report – Town Comparison: Southampton and Sheffield*. London: Reward Group.

Robinson, J. and Godbey, G. (1997) *Time for Life: The Surprising Ways Americans Use Their Time*. Pennsylvania State University Press.

Roy, C. (1991) 'Les emplois du temps dans quelques pays occidentaux', *Donnes Sociales*, 2: 223–5.

Sassen, S. (1989) 'New York city's informal economy', in A. Portes, M. Castells and L.A. Benton (eds), *The Informal Economy: Studies in Advanced and Less Developing Countries*. Baltimore, MD: Johns Hopkins University Press.

Sayer, A. (1997) 'The dialectic of culture and economy', in R. Lee and J. Wills (eds), *Geographies of Economies*. London: Arnold. pp. 16–26.

Sen, A. (1995) *Inequality Reexamined*. Cambridge, MA: Harvard University Press.

Sen, A. (1999) *Commodities and Capabilities*. Oxford: Oxford University Press.

Shucksmith, M. (2000) *Exclusive Countryside? Social Inclusion and Regeneration in Rural Areas*. York: Joseph Rowntree Foundation.

Social Exclusion Unit (1998) *Bringing Britain Together: A National Strategy for Neighbourhood Renewal*, Cm4045. London: HMSO.

Sunley, P. and Martin, R. (2000) 'The geographies of the national minimum wage', *Environment and Planning A*, 32 (10): 1735–58.

Thomas, J.J. (1992) *Informal Economic Activity*. Hemel Hempstead: Harvester Wheatsheaf.

Van Geuns, R., Mevissen, J. and Renooy, P.H. (1987) 'The spatial and sectoral diversity of the informal economy', *Tijdschrift voor Economische en Sociale Geografie*, 78: 389–98.

Williams, C.C. (2001) 'Does work pay? Spatial variations in the benefits of employment and coping abilities of the unemployed', *Geoforum*, 32: 199–214.

Williams, C.C. and Windebank, J. (1998) *Informal Employment in the Advanced Economies: Implications for Work and Welfare*. London: Routledge.

Williams, C.C. and Windebank, J. (1999) *A Helping Hand: Harnessing Self-help to Combat Social Exclusion*. York: York Publishing Services.

Williams, C.C. and Windebank, J. (2001a) 'Beyond social inclusion through employment: harnessing mutual aid as a complementary social inclusion policy', *Policy and Politics*, 29 (1): 15–28.

Williams, C.C. and Windebank, J. (2001b) 'Reconceptualising paid informal exchange: some lessons from English cities', *Environment and Planning A*, 33 (1): 121–40.

Williams, C.C. and Windebank, J. (2001c) 'Beyond profit-motivated exchange: some lessons from the study of paid informal exchange', *European Urban and Regional Studies*, 8 (1): 43–56.

Women and Geography Study Group (1997) *Feminist Geographies: Explorations in Diversity and Difference*. Harlow: Longman.

Young, M. and Wilmott, P. (1975) *The Symmetrical Family: A Study of Work and Leisure in the London Region*. Harmondsworth: Penguin.

seven

Alternative Exchange Spaces

Colin C. Williams, Theresa Aldridge
and Jane Tooke

Academics and policy-makers now widely embrace the social economy as an 'alternative' tool for combating social exclusion. To what, however, is it an alternative? And how does the answer given to this shape whether it is considered an effective alternative? Our argument in this chapter is that there are at least two ways of answering this question and that the answer heavily influences whether the social economy is considered to be an effective 'alternative' to the 'mainstream'.

The first approach is to define the economic space of the social economy as an 'alternative' to the formal sphere (Archibugi, 2000; Community Development Foundation, 1995; ECOTEC, 1998; European Commission, 1996, 1997, 1998; Fordham, 1995; OECD, 1996). In this view, which dominates discourses on the social economy, the emphasis is upon whether the social economy can be used as a means of creating formal jobs and improving employability so as to fill the gaps left by the public and private sectors. This stance is more often than not adopted by those whose normative prescription for the future of work and welfare is to return to the supposedly 'golden age' of full-employment and/or comprehensive welfare provision.

The second approach is to define the economic space of the social economy as an 'alternative' to the informal sphere. In this approach, the problem is that many people, especially the unemployed and poor, are currently unable to engage not only in the formal sphere but also in means of livelihood in the informal sphere (see Macfarlane, 1996; Williams and Windebank, 1999). Here, therefore, the emphasis is upon whether the social economy can provide those people who find themselves unable to participate in both the formal and informal spheres with alternative means of livelihood. As such, the role of the social economy is to provide access to forms of meaningful and productive activity beyond employment. This approach thus seeks to harness the social economy to compensate for the deficits of both the formal and informal spheres so that 'full-engagement', not full-employment, can be achieved.

Each of these interpretations of the role of the social economy in the future of work and welfare thus define it as an 'alternative' to very different 'mainstreams'. This chapter commences by reviewing how the social economy has been defined. We then examine how these two contrasting interpretations of the social economy define its role in tackling social exclusion and the criteria used to evaluate whether or not it is effective. We then take a case study of a prominent social economy initiative – Local Exchange and Trading Schemes (LETS) – in order to assess the contributions of the social economy in tackling social exclusion. First, we evaluate its effectiveness as an alternative to the formal economy, and secondly, we assess its effectiveness as an alternative to the informal sphere. This draws upon new empirical evidence gathered from a postal survey of all LETS co-ordinators in the United Kingdom in 1999, a national survey of LETS members and the results of in-depth, action-oriented ethnographic research with both members and non-members of LETS. In so doing, we reveal that although this social economy is relatively ineffective as an alternative to the public and private sectors, it is more effective at providing alternative means of livelihood. We then conclude by arguing that if the full potential of the social economy as an 'alternative economic space' is to be realized, there is need to move beyond evaluating it in terms of its potential to replace the private and public sectors of the formal sphere. Instead, there is a need to realize its contribution as a vehicle for collective self-help that fills the voids left by the informal sphere and promotes the 'full-engagement' of the population.

IDENTIFYING THE SOCIAL ECONOMY

The growth of interest in the social economy has brought to the fore the problematic of how to define it. These economic activities, which exist neither in the profit-maximizing parts of the market economy nor the redistributive realm of the public sector, mean that the conventional public–private and state–market dualisms that dominate thinking on how to classify economic activities no longer suffice. Instead, terms such as the 'third sector' have become increasingly popular to describe this realm of economic activity (for example, Chanan, 1999). How, therefore, can this social economy or third sector be defined?

One potentially useful approach is to appreciate the historical context of these three sectors and how their roles and functions have fluctuated over time. As Polanyi (1944) asserts, until the collapse of feudalism in western Europe, exchange between people was formed mainly in accordance with two principles: reciprocity and redistribution. He remarks that it was only after this that the market began to play any real role. But the rise of the market in England during the early nineteenth century led to strong reactions among various social groups. Workers formed trade unions so as to restrict the free supply of labour and firms formed cartels or trusts to restrict production. Thus, the free market had scarcely come into existence before attempts were made to control it. One consequence was the expansion of the public sector during the twentieth century. Even during the advent of industrialism, therefore, the newly emergent 'market' was dependent on the redistributive public sector as regulator and problem-solver. Both systems functioned side by side.

Nor, moreover, did reciprocity disappear with the rise of the market and the expansion of the public sector. Although it is often assumed that the principal site where reciprocity occurs, the family, lost its productive functions with the advent of the market, there is little evidence that this was, or indeed has been, the case. Reciprocity has persisted in realms that the public or private sector has not reached (for example, caring activities). In these realms, such exchange has complemented the private and public sectors. Indeed, the family and kinship network (and to a lesser extent neighbourhood and community networks) can still be identified as the principle carrier of the reciprocity principal (see Williams and Windebank, 1999, 2000).

However, there is also a quasi-formal realm that this principle of reciprocity inhabits. This is the 'social economy' which addresses those needs and desires that neither the private or public sectors nor the informal networks of the family, kin, neighbourhood and community, have managed to fulfil. This social economy possesses four characteristics that distinguish it from the formal private and public spheres (for example, Lorendahl, 1997; Pestoff, 1996; Westerdahl and Westlund, 1998):

- It is based on co-operative or closely-related mutual principles;
- It is based on not-for-profit principles in the sense that the initiative does not seek to expropriate a profit from its operations;
- It is private (non-public) in nature even if there is sometimes public sector involvement;
- The tasks conducted by such initiatives include economic activities that seek to fulfil people's needs and wants through the production

and/or distribution of goods and services. Put another way, it produces and sells services of a collective interest.

In addition, it possesses a characteristic that distinguishes it from more informal kinship, neighbourhood and community networks:

- Relative to associations between kin, neighbours and friends, it is a formal association that provides an organizational framework for the pursuit of collective self-help activities.

Social economy initiatives, therefore, are private formal associations for pursuing economically-oriented collective self-help based on not-for-profit and co-operative principles. They frequently occupy the voids that are filled neither by the private or public sectors, nor by the informal networks of the family, kin, neighbourhood and community. Given this definition, we can now examine the various views of its role in combating social exclusion.

THE ROLE OF THE SOCIAL ECONOMY IN TACKLING SOCIAL EXCLUSION: CONTRASTING APPROACHES

It is increasingly recognised that there are two predominant perspectives towards social exclusion (see Chanan, 1999; Jordan, 1998; Macfarlane, 1996; OECD, 1996; Robinson, 1998; Williams and Windebank, 1999). Here, each is examined in turn in terms of the way it defines social exclusion, the role that the social economy is seen to play and the criteria used to evaluate the potential of social economy initiatives.

Social economy as an alternative to the formal sphere

Based upon the idea that the only way to tackle social exclusion is to return to the 'golden age' of full-employment, this approach equates social exclusion primarily with unemployment and social inclusion principally with insertion into employment (see, for example, Jordan, 1998; Robinson, 1998; Williams and Windebank, 1999; Wilson, 1998). Social exclusion is thus approached in a very limited manner, seen predominantly in terms of exclusion from employment.

Therefore, the key issue in this approach, so far as the social economy is concerned, is whether it is capable of providing a new means of employment creation to complement the efforts of the public and private sectors. In an age of de-coupling of productivity increases from employment growth, the perception is that the private sector can no longer be relied upon to create sufficient jobs. Neither, moreover, can the post-war corporatist welfare state model be expected to spend its way out of economic problems. In this context, the social economy is seen as a potential solution. It is bolted on to existing job creation programmes and policies either as an additional means by which employment can be created beyond the public and private sectors or as a means of providing people with a springboard to enter formal employment. Indeed, the idea that the social economy can create jobs has steadily gained momentum throughout Europe and North America (European Commission, 1996, 1998; Mayer and Katz, 1985). This attachment of social economy initiatives to the goal of job generation is symbolized by the European Commission naming their new generation of projects the 'Third System and Employment' (see, for example, ECOTEC, 1998; Haughton, 1998; Westerdahl and Westlund, 1998).

To evaluate social economy initiatives, therefore, the criteria used are those associated with such initiatives, inserting people into formal employment. Evaluation criteria include the number of formal jobs created by such an initiative, its ability to facilitate skill acquisition and maintenance, whether it provides a test-bed for new potential formal businesses, its ability to develop self-esteem, and to maintain the employment ethic.

Social economy as an alternative to the informal sphere

The second approach, based on the notion of 'full-engagement', interprets social exclusion more broadly than mere unemployment. Instead, it views social exclusion as the exclusion of citizens from work (both formal and informal) and income to meet their basic material needs and creative desires (Williams and Windebank, 1999). While this approach accepts that joblessness is one important way in which people are socially excluded, it views social exclusion as a process, or a set of social relations, between those excluded and the rest of society (Alcock, 1997; Room, 1995; Wilson, 1998). As de Foucauld states:

> It is the denial or absence of social contact which fundamentally distinguishes exclusion. The dignity of the individual derives from integration in a social network – or more precisely, into a system of exchange. An individual brings something to an exchange for the other person, acquires a kind of right, recovers his [or her] status as an equal. Social exchange is what provides both a social context and autonomy which are the two essential elements of the individual. (cited in Robbins, 1994: 8)

Individuals, in other words, live in a social world based on reciprocity (see Offer, 1997). In order to take, they must be able to give. Social exclusion is therefore grounded in what people can do for others. Where reciprocity is unavailable, citizens become excluded from work and income in order to meet their basic material needs and creative desires. For social inclusion to occur, in consequence, reciprocity needs to be nurtured and re-created.

Here, therefore, the social economy is viewed from the perspective of self-help rather than job creation and its role is to provide access not only to employment but, equally, to other forms of reciprocity that might provide a means of livelihood. Social economy initiatives are thus seen as a means to stem the degradation of the social fabric in terms of the capability for reciprocal exchange (for example, Chanan, 1999; Macfarlane, 1996; OECD, 1996).

Given that the poor and unemployed lack access not only to employment but also informal reciprocal exchange (for example, Leonard, 1998; Pahl, 1984; Renooy, 1990; Thomas, 1992; Williams and Windebank, 1999), social economy initiatives are seen in this perspective to complement formal job creation initiatives. They provide a way of tackling the barriers to participation in reciprocal exchange witnessed by the poor and unemployed. These barriers are four-fold. First, they lack the money to acquire the goods and resources necessary to participate in self-help or mutual aid (economic capital). Secondly, they know few people well enough to either ask or be asked to do something (social network capital). Thirdly, they lack the appropriate skills, confidence or physical ability to engage in self-help (human capital), and fourthly, they fear being reported to the tax and/or benefit authorities if they engage in such work (an institutional barrier).

To evaluate the potential of the social economy, therefore, a wider range of indicators is employed when this approach is adopted. Besides those already listed in the first model, which indicate the ability of a social economy initiative to act as a springboard into employment, a range of other indicators is also examined which evaluate its ability to provide access to other forms of social inclusion beyond the formal labour market. These all revolve around whether the social economy initiative creates new means by which people can engage in reciprocal exchange and develop new means of livelihood outside the social relations of employment. As such, these social economy initiatives are viewed as 'complementary social inclusion policies' (CSIPs) that provide an additional means of livelihood to employment based on reciprocal exchange or collective self-help (Chanan, 1999; Macfarlane, 1996; OECD, 1996; Robinson, 1998).

Here, therefore, we evaluate the effectiveness of a prominent social economy initiative, namely Local Exchange and Trading Schemes (LETS), as a tool for combating social exclusion. To do this, we evaluate its effectiveness in terms of first, its ability to create formal jobs and thus bring about 'full-employment', and secondly, its ability to harness self-help activity and thus facilitate 'full-engagement'. This will then enable us to draw conclusions about the role of social economy initiatives.

EVALUATING LETS AS A MEANS OF TACKLING SOCIAL EXCLUSION

A LETS is created where a group of people form an association and create a local unit of exchange. Members then list their offers of, and requests for, goods and services in a directory that they exchange, priced in a local unit of currency. Individuals decide what they want to trade, who they want to trade with, and how much trade they wish to engage in. The price is agreed between the buyer and seller. The association keeps a record of the transactions by means of a system of cheques written in the local LETS units. Every time a transaction is made, these cheques are sent to the treasurer, who works in a similar manner to a bank sending out regular statements of account to the members. No actual cash is issued since all transactions are by cheque and no interest is charged or paid. The level of LETS units exchanged is thus entirely dependent upon the extent of trading undertaken. Neither does one need to earn money before one can spend it. Credit is freely available and interest-free.

As such, LETS are very much a social economy initiative. They are private, formal associations for pursuing economically-oriented collective self-help, based on not-for-profit and co-operative principles. They operate in order to fill the voids existing in the provision of needs and wants that are fulfilled neither by the private or public sectors, nor by the informal networks of the family, kin, neighbourhood and community. These initiatives are currently one of the most prominent social economy initiatives, being advocated in the UK. Whichever government document is examined, LETS are often the principal social economy initiative, heralded as a potential solution to social exclusion. In the UK, for instance, hardly a government report on social exclusion passes without some mention of these local currency schemes (DETR, 1998; DSS, 1999; Home Office, 1999; Social Exclusion Unit, 1998, 2000).

Indeed, by 1999, LETS had spread to many of the advanced economies. There were approximately 303 LETS in the UK, 300 in

France, 250 in Australia, 110 in the USA, 100 in Italy, 90 in Holland, 90 in Germany, 57 in New Zealand, 29 in Belgium, 27 in Canada, 19 in Austria, 1 in Switzerland, 14 in Sweden, 7 in Norway and 3 in Denmark, to name but a few nations. Up until now, however, there has been no comprehensive attempt to evaluate their effectiveness as a means of tackling social exclusion or to offer lessons on how this could be further improved. Instead, only one-off studies of individual LETS have been conducted (for example, North, 1998, 1999; Pacione, 1997a, 1997b; Seyfang, 1998; Williams, 1996a, 1996b, 1996c). Given the frequently contrasting methods used, not only were these studies often not comparable, but one did not know whether the findings were unique to the individual LETS or of wider significance.

To evaluate the effectiveness of LETS at both helping members into formal jobs and providing complementary means of livelihood, three methods were used during 1998 and 1999. First, a survey of all LETS co-ordinators was undertaken in 1999. These co-ordinators were identified from existing data sets and, additionally, by 'snow-balling' methods. Of the 303 LETS identified and surveyed, 113 responded (37 per cent). Secondly, a membership survey was conducted. From the results of the co-ordinators' survey, maximum variation sampling was used to identify widely different types of LETS in existence (for example, by membership size, urban/rural location, size of area covered, type of members, time established). Surveying 26 LETS, 2,515 questionnaires were sent out and 810 (34 per cent) returned. Finally, in-depth, action-oriented ethnographic research was conducted on two LETS in very different locations: the semi-rural area of Stroud and the deprived urban area of Brixton in London. Here, we report the results of these surveys along with the qualitative data from Stroud.[1]

Starting with the overall magnitude of LETS, the co-ordinators' survey identified that the LETS responding had an average membership of 71.5 members and an average turnover equivalent to £4,664. If the responding LETS are taken as representative of all LETS in the UK, then the total UK LETS membership is the equivalent of 21,816 and the total turnover equivalent to some £1.4 million. In terms of the total exchange-value of LETS, therefore, these schemes are relatively insignificant. However, when measured in terms of their use-value in tackling social exclusion, as we shall show below, they become more effective vehicles.

Who, therefore, joins LETS? Of the 810 members responding to the survey, LETS members are predominantly aged 30–49, women, from relatively low income groups and those who are either not employed or are self-employed. Hence, it appears that the membership profile of LETS is skewed towards the socially excluded, if one accepts that non-employment and low incomes are (surrogate) indicators of social exclusion.

Why do these people join LETS? Just 2.4 per cent join explicitly to use it as a means of gaining access to employment and these are all people seeking to promote or gain business so as to become self-employed. The rest join either for ideological reasons (23.3 per cent) or to engage in complementary means of livelihood for economic and community-building reasons (74.3 per cent). Economic reasons are cited predominantly by the relatively poor and unemployed, who state demand-side rationales such as their lack of money to buy goods and services formally (12 per cent of members), their desire for a specific service (9.3 per cent) or their wish to exchange goods and services (20.8 per cent). Only 1.3 per cent of members state a supply-side rationale and this was always to use their skills. Meanwhile, 22.9 per cent of all members, who are near enough entirely the employed and relatively affluent, cite community-building rationales.

Given this concentration of low-income households and non-employed people in the membership and the fact that they join to develop complementary means of livelihood for economic reasons, we now evaluate the effectiveness of LETS, first, as springboards into employment, and second, as complementary means of livelihood.

EVALUATING LETS AS AN ALTERNATIVE TO THE FORMAL SPHERE

To evaluate LETS in this regard, one must examine the number of formal jobs created, their ability to facilitate skills acquisition and maintenance, whether they provide a test-bed for new potential formal businesses, and their ability to develop self-esteem and to maintain the employment ethic.

Examining the number of direct jobs created by LETS, the finding is that this amounts to no more than a handful since mostly volunteers run them. Nevertheless, some 4.9 per cent of members surveyed asserted that LETS had helped them gain formal employment. This was entirely because working in the LETS office administering the scheme had enabled valuable administrative skills to be acquired which they had then been able to use to apply successfully for formal jobs. In no other way, however, did the LETS enable participants to use them as a springboard to become an employee. As such, their ability in this regard is limited since only a small number of people can at any one time play a prominent role in administering the scheme.

For others, however, LETS represented a springboard into self-employment. Some 10.7 per cent of respondents asserted that LETS had helped them become self-employed by enabling them to develop their client base (cited by 41.2 per cent of those who were self-employed), easing the cash-flow of their business (cited by 28.6 per cent) and enabling them to use it as a test-bed for their products and services, cited by nearly all who defined themselves as self-employed. As several respondents stated during interviews and/or focus group discussions:

> I was looking to start off as a freelance journalist at the time, and ... [joining LETS] was just another way of generating some work and some contacts and building up experience without having to put in, sort of, the risk of hard currency. (man aged 35–39 in focus group discussion)

> We joined as a way of getting into doing things on quite a small scale without having to have this big risk thing of going into it as a small business so I make things – arts and crafts stuff – which I can sell through the LETS and sort of get an idea of what people actually like. I found it really useful as a way of getting back into making things again, and it really does boost your confidence being able to sell your stuff. If I'm selling it for money a lot of people don't have that much excess money to spend on stuff, but they've got a lot of excess LETS, so yeah they can buy your things, it's like, 'Ooh someone wants to buy my stuff, it must actually be alright, so you sort of go back and make more knowing that it is actually okay. (single woman aged 30–34 in focus group discussion)

> I became a LETS member and used the LETS as a source to advertise my services and from this I have managed to go self-employed. All of my customers are coming through the LETS and my business is slowly building up. The LETS has been extremely important in this development both financially and in the community support it provides – I get my childcare paid for through the LETS which enables my business development. LETS has enabled my survival. At the moment life is very tight, I'd be desperate without LETS. (self-employed, female single parent aged 35–39 on Income Support who set herself up as a self-employed massage therapist and has transferred to Family Credit).

More indirectly, some 24.8 per cent of all respondents asserted that the LETS had boosted their self-confidence. This was particularly the case among younger people, the registered unemployed and those with few qualifications. In addition, 15.9 per cent that it had enabled new skills to be acquired (24.3 per cent of the registered unemployed), mostly related to computing, administration and interpersonal skills. By recognizing and valuing work beyond employment, therefore, LETS were not only helping members gain entry to employment or develop self-employed business ventures, but also providing them with greater personal transferable skills and self-confidence which would later be of use to them in the formal labour market. As a 50–54 year-old unemployed single woman put it:

> Coming into LETS I've had a lot of interaction with other people, lots of different people, and it helps me with my confidence. I'm going to learn how to do the directory, and I've been inputting cheques into the computer accounts so I'm learning different things through my LETS work. I think I just enjoy the contact with other people and the fact that I'm getting LETS responsibilities now. It makes me feel that I'm a bit important and getting invited to meetings, it's really good. And writing up messages in the day book, someone put, 'Good idea, well done!' Well, it just makes you feel valued and that you are making a contribution. ... I've been out of work for over two years and I've had problems getting references from previous employers because they say that they can't remember that long ago, which is upsetting ... so I should be able to get references from the LETS for the work I'm doing, which will help in looking for paid work when I'm ready.

LETS, in consequence, do appear to be a useful springboard into employment for a small but significant proportion of members. Within the logic of this perspective, therefore, several possible policy responses arise. To further develop this social economy initiative as a generator of employment one could first fund LETS office workers for their administrative work, such as under the 'voluntary and community' sector of New Deal. This would provide workers with a proven means of entering the formal labour market as employees and, at the same time, enable the more efficient running of the LETS (since it would not be so reliant on volunteers for its day-to-day administration). Then, many who are currently operating as self-employed in LETS could be encouraged to enter the 'self-employment' option in New Deal, and their trading on LETS could be recognized as part of their attempt to become self-employed. As we shall see later, however, such policy responses would be strongly resisted by many LETS members, who contest both the view of the social economy underlying such initiatives as well as its potential impacts on LETS.

EVALUATING LETS AS AN ALTERNATIVE TO THE INFORMAL SPHERE

In order to evaluate the effectiveness of LETS in complementing the informal sphere, the extent to which they counter the barriers discussed above need to be analysed. Starting with the extent to which LETS tackle the barrier of economic capital, some 40 per cent of members assert that LETS provided them with access to interest-free credit (rising to 62.1 per cent of the registered unemployed and 51 per cent of low-income households). LETS therefore provide people with access to money. For two-thirds (64.5 per cent) of the registered unemployed, this had helped them cope with unemployment, with some 3.1 per cent of their total income coming from their LETS activity.

LETS also enable their participants to tackle the barrier of social network capital. Some 76.2 per cent of respondents assert that the LETS helped them to develop a network of people upon whom they could call for help, while 55.6 per cent assert that it has helped them develop a wider network of friends and 31.2 per cent deeper friendships. LETS therefore develop 'bridges' (that is, bringing people together who did not know each other before) more than 'bonds' (that is, bringing people who already know each other closer together). Given that most members lacked kinship networks in the localities they inhabited and that kinship networks are the principal source of mutual aid in contemporary society (Williams and Windebank, 1999), LETS thus provide those without such a local network with a substitute. Some 95.3 per cent of LETS members had no grandparents living in the area, 79.5 per cent no parents, 84.3 per cent no brothers or sisters, 58.2 per cent no children, 92.6 per cent no uncles or aunts, and 90.8 per cent no cousins.

This important role that LETS play in developing social networks was brought out in numerous interviews and focus group discussions. Take, for example, the following extract from a focus group discussion:

> Discussant 1: We joined LETS only a year ago, but moving to a new area you don't have your family and friends readily laid on, and it's a very good way to get to know people. ... If you haven't got anybody or you don't know your neighbours very well, then it's a great way of asking people to do something a bit silly that you wouldn't be able to do. The first person we contacted, we wanted something moving and we couldn't lift it ourselves and we thought oh, we've got no neighbours, or they're old neighbours, so it was sort of the introduction to LETS. (part-time employed woman aged 30–34)
>
> Discussant 2: We didn't actually know anyone to help us carry something. (full-time employed woman aged 30–34)
>
> Discussant 1: So it was as simple as that, so that was the starting point and now we're just looking to get into debt and spend more LETS and get involved that way. (as above)

As such, LETS is seen as a way of creating mostly 'bridging' social capital (that is, bridges between people who did not previously know each other). However, for some people, especially unemployed members who would otherwise have relatively few opportunities to forge new social networks, it is also being used to develop 'bonding' social capital (that is, closer bonds between people who before loosely knew each other). As an unemployed single woman aged 50–54 stated during an interview:

> When I first moved here, I was finding it very, very hard to meet people, make friends, people are very reserved around this area, they don't sort of welcome strangers with open arms; so I thought I could meet people through the LETS system, and that's worked really well ... you see that's the one thing about being on benefits, low-income, it's exceedingly useful (LETS), it's a way of instead of barely existing, you know it enables you to do a lot of things, and its good, and like I was very isolated when I first moved here and through the LETS I don't feel so isolated at all now, I've got lots of people I know to speak to, and I've got a couple of very good friends, it's great, there's a network of people available.

Besides tackling the barriers of economic and social network capital, there is also evidence that LETS tackle the barrier of human capital that can constrain participation in reciprocity. As discussed above, LETS provide an opportunity to both maintain and develop members' skills as well as to rebuild their self-confidence and self-esteem by engaging in meaningful and productive activity that is valued and recognized by others, who display a willingness to pay for such endeavour. Finally, there is an institutional barrier to pursuing complementary means of livelihood. Many who are unemployed are fearful of being reported to the authorities, even if they engage in unpaid mutual aid. This is not currently being overcome by LETS. Although only 13 per cent of members feel worried about tax liabilities and 12 per cent about reductions in welfare payments, 65 per cent of registered unemployed members are concerned about their situation. Similarly, and as identified in a survey of 103 non-members in Stroud, grave concerns exist among unemployed non-members that their trades will result in a reduction in benefits and this makes them wary of joining. Ironically, therefore, those who would most benefit from LETS are discouraged from joining and trading due to the uncertainty over their legal position *vis-à-vis* benefits. The current *laissez-faire* approach of government, in consequence, is insufficient to appease both members and non-members who are registered unemployed.

CONCLUSIONS

To examine the role of the social economy, this chapter has evaluated the potential of LETS. It has revealed that although this prominent social economy initiative appears to be successful for some as a springboard into employment, it has significance for a wider proportion of members as a vehicle for providing complementary means of livelihood. Consequently, if social exclusion is equated solely with unemployment and the social economy is viewed merely as a potential springboard into employment for the unemployed, then the full value of such social economy initiatives will not be recognized. However, if social exclusion is more broadly conceptualized as the exclusion of citizens from work (both formal and informal) and income to meet their basic material needs and creative desires, then the broader significance of such social economy initiatives can be realized.

In other words, this analysis has uncovered that assessing such initiatives in terms of their ability to fill the gaps left by the public and private sectors of the formal sphere fails to recognize the full value of social economy initiatives. The principal contribution of social economy initiatives such as LETS is that they provide members with the opportunity to engage in meaningful and productive activity, develop their human capital and get work done which they could not otherwise afford if they had to pay formally. It also provides them with the social networks to enable them to seek out somebody both to help and be helped by. They are a lot less effective, however, at creating formal jobs. As a tool for achieving 'full-employment', therefore, LETS are relatively ineffective. However, as a vehicle for facilitating 'full-engagement', LETS have been shown to be a useful tool.

The current problem, however, is that policies towards social economy initiatives such as LETS are predicated on the premise of full-employment where social inclusion is equated with employment. The first change that is required, therefore, is that it needs to be recognized that the value of such social economy initiatives lies in their potential to harness self-help, not create formal jobs. In other words, there needs to be shift from an employment-ethic that views formal jobs as the only means of social inclusion to a 'work'-ethic that recognizes other avenues beyond employment by which people can seek inclusion.

Until this is achieved, then, this social economy initiative will be unable to fulfil its potential. What is certain from this evaluation, however, is that the hegemonic totalizing discourse of full-employment as the only route out of poverty limits the possibilities and opportunities of the social economy. Until it is cast off, policy will be unable to start to look forward to alternative ways of organizing work and welfare. Indeed, there is much to be learnt from LETS members themselves in this regard – they are using this social economy initiative as a vehicle for providing themselves with complementary means of social inclusion beyond employment rather than merely as a vehicle for helping them into formal employment. It is to be hoped that policy-making can learn from the actions of these pioneers of a new approach towards revitalizing communities.

Note

The authors would like to thank the Economic and Social Research Council (ESRC) for funding this research project (R000237208). We would also like to thank all of the national LETS co-ordinators and members who responded to the surveys as well as the membership of Stroud and Brixton LETS for being so open with us during the eight-month visit of the researcher in each location.

1. Between November 1998 and April 1999, in-depth action research (see Stringer, 1996) was conducted with this LETS. The first stage comprised an initial survey of the 326 members to identify the character of the membership, trading levels and members' perceptions of the effectiveness of LETS in promoting social inclusion. The second stage used a multi-method approach, including in-depth interviewing, focus groups and the researchers' participation in all aspects of the scheme to explore the multiple meanings of participation, visions of LETS development, and the barriers to increased participation. In total, 29 in-depth interviews were completed and transcribed, with transcripts being returned to the interviewee for final edit and agreement for use in this research. Five focus groups were moderated with numbers varying from between four and seven participants (excluding the facilitator). Again, these were fully transcribed and returned to participants for final edit, and this process was evaluated by the participants.

References

Alcock, P. (1997) *Understanding Poverty*. London: Macmillan.

Archibugi, F. (2000) *The Associative Economy: Insights beyond the Welfare State and into Post-Capitalism*. London: Macmillan.

Chanan, G. (1999) 'Employment and the social economy: promise and misconceptions', *Local Economy*, 13 (4): 361–8.

Community Development Foundation (1995) *Added Value and Changing Values: Community Involvement in Urban Regeneration: A 12-Country Study for the European Union*. Final report CEC DG XVI. London: Community Development Foundation.

DETR (1998) *Community-based Regeneration Initiatives: A Working Paper*. London: Department of the Environment, Transport and the Regions.

DSS (1999) *Opportunity for All: Tackling Poverty and Social Exclusion*, Cmnd 4445. London: HMSO.

ECOTEC (1998) *Third System and Employment: Evaluation Inception Report*. Birmingham: ECOTEC.

European Commission (1996) *Social and Economic Inclusion through Regional Development: The Community Economic Development Priority in ESF Programmes in Great Britain*. Brussels: European Commission.

European Commission (1997) *Towards an Urban Agenda in the European Union*. Communication from the European Commission COM (97)197 final report. Brussels: European Commission.

European Commission (1998) *The Era of Tailor-made Jobs: Second Report on Local Development and Employment Initiatives*. Brussels: European Commission.

Fordham, G. (1995) *Made to Last: Creating Sustainable Neighbourhood and Estate Regeneration*. York: Joseph Rowntree Foundation.

Haughton, G. (1998) 'Principles and practice of community economic development', *Regional Studies*, 32 (9): 872–8.

Home Office (1999) *Community Self-help*. London: Home Office.

Jordan, B. (1998) *The New Politics of Welfare: Social Justice in a Global Context*. London: Sage.

Leonard, M. (1998) *Invisible Work, Invisible Workers: The Informal Economy in Europe and the US*. Basingstoke: Macmillan.

Lorendahl, B. (1997) 'Integrating public and co-operative/social economy: towards a new Swedish model', *Annals of Public and Co-operative Economics*, 68 (3): 379–96.

Macfarlane, R. (1996) *Unshackling the Poor: A Complementary Approach to Local Economic Development*. York: Joseph Rowntree Foundation.

Mayer, M. and Katz, S. (1985) 'Gimme shelter: self-help housing struggles within and against the state in New York City and West Berlin', *International Journal of Urban and Regional Research*, 9 (1):123–56.

North, P. (1998) 'Exploring the politics of social movements through "sociological intervention": a case study of Local Exchange Trading Schemes', *The Sociological Review*, 46 (3): 564–82.

North, P. (1999) 'Explorations in heterotopia: Local Exchange Trading Schemes (LETS) and the micro-politics of money and livelihood', *Environment and Planning D: Society and Space*, 17 (2): 69–86.

OECD (1996) *Reconciling Economy and Society: Towards a Plural Economy*. Paris: OECD.

Offer, A. (1997) 'Between the gift and the market: the economy of regard', *Economic History Review*, 2: 450–76.

Pacione, M. (1997a) 'Local Exchange Trading Systems as a response to the globalisation of capitalism', *Urban Studies*, 34: 1179–99.

Pacione, M. (1997b) 'Local Exchange Trading Systems: a rural response to the globalisation of capitalism?', *Journal of Rural Studies*, 13 (4): 415–27.

Pahl, R.E. (1984) *Divisions of Labour*. Oxford: Basil Blackwell.

Pestoff, V.A. (1996) 'Work environment and social enterprises in Sweden', paper presented to the European Conference on Labour Markets, Unemployment and Co-ops in Budapest, 27–28 October.
Polanyi, K. (1944) *The Great Transformation*. Boston, MA: Beacon Press.
Renooy, P. (1990) *The Informal Economy: Meaning, Measurement and Social Significance*. Netherlands Geographical Studies, No. 115. Amsterdam: Netherlands Geographical Institute.
Robbins, D. (1994) *Social Europe. Towards a Europe of Solidarity: Combating Social Exclusion*. Brussels: European Commission.
Robinson, P. (1998) 'Employment and social inclusion', in C. Oppenheim (ed.), *An Inclusive Society: Strategies for Tackling Poverty*. London: IPPR.
Room, G. (ed.) (1995) *Beyond the Threshold: The Measurement and Analysis of Social Exclusion*. Cambridge: The Policy Press.
Seyfang, G. (1998) 'Green money from the grassroots: local exchange trading schemes and sustainable development'. Submitted PhD thesis. Leeds: Leeds Metropolitan University.
Social Exclusion Unit (1998) *Bringing Britain Together: A National Strategy for Neighbourhood Renewal*. London: HM Cabinet Office.
Social Exclusion Unit (2000) *National Strategy for Neighbourhood Renewal: A Framework for Consultation*. London: HM Cabinet Office.
Stringer, E.T. (1996) *Action Research: A Handbook for Practitioners*. London: Sage.
Thomas, J.J. (1992) *Informal Economic Activity*. Hemel Hempstead: Harvester Wheatsheaf.
Westerdahl, S. and Westlund, H. (1998) 'Social economy and new jobs: a summary of twenty case studies in European regions', *Annals of Public and Co-operative Economics*, 69 (2): 193–218.
Williams, C.C. (1996a) 'Local Exchange and Trading Systems (LETS): a new form of work and credit for the poor and unemployed', *Environment and Planning A*, 28 (8): 1395–415.
Williams, C.C. (1996b) 'The new barter economy: an appraisal of Local Exchange and Trading Systems (LETS)', *Journal of Public Policy*, 16 (1): 55–71.
Williams, C.C. (1996c) 'Informal sector responses to unemployment: an evaluation of the potential of Local Exchange and Trading Systems (LETS)', *Work, Employment and Society*, 10 (2): 341–59.
Williams, C.C. and Windebank, J. (1999) *A Helping Hand: Harnessing Self-help to Combat Social Exclusion*. York: Joseph Rowntree Foundation.
Williams, C.C. and Windebank, J. (2000) 'Helping each other out? Community exchange in deprived neighbourhoods', *Community Development Journal*, 35 (2): 146–56.
Wilson, R. (1998) 'Comment: citizen's involvement', in C. Oppenheim (ed.), *An Inclusive Society: Strategies for Tackling Poverty*. London: IPPR.

eight

Alternative Lifestyle Spaces

Jeffrey Jacob

Although the USA at the beginning of the twenty-first century is an overwhelmingly urban nation (at the end of the twentieth century only 2.2 per cent of the population lived on farms (Albrecht and Murdock, 1990)), its agrarian past periodically reasserts itself. Politicians, when it suits their purposes, evoke frontier values of self-reliance and community resourcefulness. Advertisers draw on both the sentimentalism and the rugged individualism of the countryside to sell their wares, from lemonade to pick-up trucks. On occasion, television series with rural themes like *Little House on the Prairie* or *The Waltons* become cultural institutions during both their prime time and syndication runs. Much of classic cinema, with movies like *Gone with the Wind* and *The Grapes of Wrath*, employ rural imagery. Then there is Dorothy in the *The Wizard of Oz*. As a secularized myth of the American experience, *The Wizard of Oz* has the Kansas farm girl Dorothy in Oz journeying down the Yellow Brick Road towards the glitter of the Emerald City – 'most of it was beautiful', she later tells Auntie Em. But though she is attracted by the excitement of the Emerald City, Dorothy in the end chooses the black and white landscape of her Kansas farm home. 'All I kept saying to everybody was "I want to go home".' Clicking her ruby slippers together, and repeating three times, 'There's no place like home', Dorothy finds herself back in the familiar surroundings of a Kansas farm house, with the last words of the script includes the refrain: 'And … oh Auntie Em, there's no place like home' (Langley et. al., 1989 [1939]: 132).[1]

Unlike Dorothy, however, rural and small-town residents throughout the twentieth century, America's urban century, have left the countryside to seek their fortunes in the country's central cities and suburbs. Nevertheless, over the course of the century there has also been a steady, though fluctuating, counter-stream migration back to rural America. This 'turnaround migration', in contrast to rural-to-urban migration, is often motivated by non-economic factors, including opportunities to be close to family and friends, to enjoy the natural beauty of the outdoors, and to experience the intimacy of small-town life.[2] Important segments of the counter-urbanization trends have been the back-to-the-land movements of the early and latter part of the twentieth century (Jacob, 1997). The back-to-the-land movement is a diverse phenomenon that primarily consists of former urban residents taking up the practice of semi-subsistence agriculture on small plots of marginal farmland. One student of the late twentieth-century back-to-the-country experience estimated that there were over one million neo-homesteaders in rural America at the height of the movement at the end of the 1970s.[3]

While the back-to-the-land movement constitutes a genuinely alternative lifestyle, few members of the movement, with the exception of the less than 5 per cent from the survey reported on below who run commercial micro-farms or live on relatively self-sufficient homesteads, are able to carve out alternative economic space in terms of full-time employment. At the same, the back-to-the-landers surveyed say, on average, that they produce approximately 30 per cent of their food needs from their properties, not an inconsiderable contribution in the direction of alternative economic space. But most have to find work off their smallholdings in order to survive economically and pursue the perceived pleasures of country living.

While back-to-the-landers currently reside on the margins of the agricultural economy, this chapter takes the position that the neo-yeomen possess the potential to become major players in a restructured agricultural marketplace. In order to make a case for the possible revitalization of yeoman agriculture, the chapter locates the late twentieth-century back-to-the-land movement within the contexts of both agrarian history and contemporary farm policy debates. Global agribusiness in the form of factory farming has all but suffocated the family farming tradition, casting the dreamers who head back-to-the-land as eccentrics who support their country habits with non-farm income. Nevertheless, through the first half of the nineteenth century small-scale, labour-intensive farms fed America. And although the early twenty-first century is radically different from the mid-nineteenth century, there are agrarian philosophers and policy analysts who see human-scale, intensive agriculture as a natural corrective to factory farming and its devastation of communities and their natural environments.[4]

But even if agricultural policy were at some point in the foreseeable future able to engineer the dismantling of global agribusiness, could back-to-the-landers, as neo-yeoman, enter a restructured marketplace as serious producers to help feed an overwhelmingly urban nation, as their spiritual ancestors, the nineteenth-century yeoman, once did? The answer to this question begins with an understanding of the dynamics of yeoman America, and how over the course of 200 years they came to be replaced by corporate agriculture. Central to this historical overview is the contention that yeomen farmers and their late nineteenth-century family farming decendents were not replaced by factory farming because of their inefficiencies or that they were incapable of producing the foodstuffs for urban American. The historical record indicates just the opposite: family farmers lost their land because they were too efficient. American agricultural history is the story of chronic surpluses driving down commodity prices to the point where individual producers are unable to recover their costs or earn a fair living. Corporate farmers, who have made significant contributions to the surpluses, move in once the family farmers leave (see Greider, 2000).

After the review of the rise and demise of yeoman America, the chapter continues with a description of the twentieth-century back-to-the-land movements as a reaction to the eclipse of the yeoman tradition, and then concludes with a consideration of the policy options that could bring neo-yeoman and family farmers back to the centre of agricultural production. Intriguingly, and even ironically, the revival of the yeoman tradition revolves around the possibility of returning American agriculture to market principles. Historically, yeomen were victims, with their chronic surpluses, of market forces. But a social market approach which demands all costs in the agricultural production and distribution equation be reflected in commodity prices, as opposed to the nominally free market which ignores social and environmental costs, could very well reconfirm the efficiency of yeoman and neo-yeoman agriculture, and in the process marginalize factory farming.

YEOMAN AMERICA

Back-to-the-land is a pilgrimage in search of the yeoman farm family of early republican America just as much as it is an expedition across contemporary interstate highways and back roads to find one's own smallholding. Looking across 200 years through a lens clouded by nostalgia, we attempt to locate the noble yeomen, stewards of the earth, whom Thomas Jefferson saw as 'the chosen people of God, if ever He had a chosen people, whose breasts He made His peculiar deposit for a substantial and genuine virtue' (Shi, 1985: 77–8).

Historians have pursued yeoman farmers through probate proceedings, account books, letters and diaries, and have not been able to agree completely on their character or the nature of the world they inhabited.[5] Pulling together fragments of yeoman culture left in the historical record, and allowing scepticism to penetrate legend, it is still possible to construct a coherent picture of small-farm life in the early nineteenth century, a way of living neo-agrarians are trying to recreate at the end of the twentieth century.

The material basis of the yeoman way of life was subsistence agriculture. The defining characteristic of subsistence agriculture is the union of production and consumption. At the level of the individual farm, family members worked together to plant and harvest grain, tend vegetable gardens and care for livestock. The farm home was the location for a variety of assembly and manufacturing operations. Women turned wool and flax into cloth and then transformed it to the homespun attire worn by the common folk in pre-industrial America.

But even though farm families produced most of what they consumed, they were not self-sufficient. Survival in yeoman society was a community enterprise. Individual families specialized in a variety of essential activities, from blacksmithing and tanning to gristmilling and storekeeping, and then traded goods and services through neighbourhood exchange networks. Money was a rare medium of exchange in these trading networks. 'Money is so scarce and hard to be got [that] we must live without buying much', was the way an Illinois woman in the 1830s characterized her family's situation in a letter to a sister in Ohio, but then added, 'most people get along on what they produce and trade with others for' (Faragher, 1985: 178).

Family production and the community exchange networks could not provide everything the yeoman families needed, or wanted. If nothing else, they needed cash to pay taxes. In addition, currency was usually required for payment on their ubiquitous mortgages. And while merchants would often accept local produce for salt, sugar, window glass and gunpowder, long-term trading relationships with storekeepers required some cash. The necessary cash came by way of ventures into commercial commodity markets. Families would sell the surplus portions of their harvests to acquire the cash needed to round out their subsistence way of life. It might, then, be more accurate to label yeoman agriculture as *semi*-subsistent, rather than as simply subsistence-based. The surplus grain from hundreds of thousands of smallholders constituted a substantial portion of the foodstuffs circulating in early American regional, national and international commodity markets.

Practising their unique brand of subsistence agriculture on the margins of the market economy, the yeoman farm families developed a distinct agrarian culture. It was certainly American, but by the standards of today's preoccupations with productivity and efficiency, the culture of subsistence agriculture is quaintly anachronistic. Yeoman values emphasized co-operation and community welfare and downplayed competition and acquisitiveness. Communal values are hardly foreign to modern America; yeoman culture, though, practised a relaxed sociability that would make the self-driven representative of post-industrial culture uncomfortable. In the hierarchy of the yeoman's worldview, personal relationships took precedence over individual accumulation. An ethic of 'enoughness' insulated camp meetings, church socials and neighbourly visiting from preoccupation with accumulation.[6]

The romantic edge can be taken off the description of the yeoman way of life by seeing yeoman values rooted in the nature of semi-subsistence agriculture itself, rather than as a consequence of spontaneous virtue. Once families had grown food for their own basic needs and provided the clothing and shelter to protect themselves from the elements, there was a definable end to their work. Since they needed only a small production surplus to earn cash and since goods and services the individual family could not supply for itself came by way of community exchange networks, there was little incentive to strive for more productivity and efficiency. Storage technology was not sophisticated, and the cost of transportation could exceed the value of produce going to market. Relieved then from the pressure of surplus production, yeoman farmers turned their attention to cultivating personal relationships within their communities, the ultimate source of their well-being.

The outside world, of course, did not allow the yeoman to retreat to tranquil community life. The outside world offered an invitation to progress. Through stepped-up participation in commercial commodity markets yeomen could bury backwardness and exchange frugality for affluence. Many yeomen, however, were reluctant to accept the first invitations to work for the market. They had their own priorities – a quality of life that could not be quantified on an accountant's ledger. In addition, they harboured a profound animus for the commercial culture of the eastern seaboard. Its arrogance and luxury repelled them. They saw themselves of an agrarian tradition whose labour supported the unproductiveness of the merchant elite.

To the yeomen, the commercial market was abstract and artificial. They were accustomed to the face-to-face trading relationships of the local community, rather than dealing through intermediaries who quoted seaport prices. Value was very much a function of physical labour, whether the product was a bushel of wheat or a straight-backed chair. Prices that fluctuated with supply and demand both mystified and frightened them. In the elaborate trading networks of their own neighbourhoods, exchange values remained constant. Theirs was a concrete, predictable world; the market promised wealth, but left an endemic uncertainty.

Yeoman reservations about the market were more than theoretical. Banks foreclosed on their mortgages; they spent years in debtor prisons for unpaid notes; and the courts were unsympathetic to their calls for relief. 'I just got tired of working for the other fellow', was the way one encumbered and delinquent yeoman put it, and then added, 'I worked and toiled from year to year and the fruits of my labor went to the man who never struck a lick' (Hahn, 1985: 195).

Back-country farmers practised safety-first agriculture. They avoided risk and sought security. The market undermined their cautious approach to making a living. Market prices that could fluctuate wildly made inevitable debt for land and other necessities a dangerous gamble. The 1819 collapse in world commodity prices was a shock that closed down the game for thousands of independent yeoman families and left them as tenant farmers. Wheat prices, for example, from 1817 to 1821 fell from $2.41 a bushel to 88¢; cotton went from 33¢ a pound to 14¢; and tobacco from 40¢ to 4¢ (Sellers, 1991: 135).

This kind of market experience left the yeoman farmers deeply suspicious of the whole complex of commercial institutions: credit, banks, courts, limited liability corporations, lawyers, and money itself. They had to have money to pay taxes and mortgages, but having to make it undercut their independence. Courts and legislators in turn did not always sympathize with the honest yeoman in default. Members of the commercial establishment were oblivious to the yeoman's predicament because they saw the instruments of the yeoman's oppression, banks and credit, as the necessary engines of growth and development. The smallholders profoundly frustrated the east-coast elites, who believed the yeomen to be backward, civically irresponsible and ultimately dangerous to their vision of national development.[7]

A constellation of forces eroded the yeoman's resistance to the commercial marketplace. In the second quarter of the eighteenth century canals and turnpikes, and in the nineteenth century paddle-wheel steamboats and railroads, all carved arteries deep into America's interior.[8] Small-farm families could sell their wheat or tobacco on commodity markets without having shipping costs take away their profits. Mortgages and accumulated debts with local merchants made taking advantage of these new trading opportunities all the more tempting. The new agricultural technology of iron and steel plows and mechanical reapers, mowers and threshers, and the general application of horse-power to agriculture, as the yeoman 'learned to farm sitting down' (Bogue, 1963), increased the smallholders' productivity and the size of surplus available for the market.

The deciding factor, though, in breaking down yeoman resistance to the allure of the market was not so much a multitude of outside forces, but an integral characteristic of the farm family itself.[9] Subsistence agriculture required a large contingent of unpaid family members to perform the day-to-day labour necessary to maintain the smallholder way of life. As a family of eight to ten children grew to maturity there was a tremendous pressure to provide each child with an inheritance – and independent, productive children would also ensure the parents security in their declining years. In an agrarian society, an inheritance meant land and, consequently, smallholders had to be preoccupied with working not only their own land, but pulling together enough property to keep the extended family together. Just about the only way, however, to buy more land was by selling increasingly larger shares of surplus production in the commercial commodity markets. The market of course could pay off handsomely in good years. Bad years often brought a cruel end to yeoman dreams. Without the cash to pay off the debts of expansion, smallholders could lose everything, caught in a foreclosure cycle whose discords have been a refrain heard with regularity over the course of the history of rural America.

THE ECLIPSE OF THE YEOMAN TRADITION

Despite the outside pressures and the internal strain, yeoman farming remained a dominant tradition in American agriculture through the middle of the nineteenth century. The Civil War, however, precipitated a dramatic change in the face of rural America. The process of change was to assign eccentricity status to subsistence agriculture, leaving it resident only in folklore by the start of the Great Depression. The war's disruption of conventional agricultural supply lines and its demand for food to feed armies who would otherwise be working their own farms, created a market pull that smallholders were unable to resist. Then the agricultural boom years continued after the war, as immigrants and high fertility rates more than doubled and tripled the size of cities like New York, Philadelphia and Chicago by the start of the twentieth century.[10]

In rising to the challenge of feeding urban America, rural America was to find failure in its very success. It simply did the job too well. The country's population would rise by 25 per cent between 1870 and 1880, a considerable increase, but during the same period agricultural production soared a phenomenal 53 per cent. This kind of mismatch between supply and demand precipitated a depression in agricultural prices that lasted from 1870 to 1900. The percentage of the labour force in agriculture dropped from 60 per cent in 1860 to 40 per cent in 1900. A wave of foreclosures made one in four farmers tenants on the land they tilled, a figure that was to increase to nearly four in ten by 1910. In addition, there were 4.5 million agricultural day-labourers in rural America by the end of the nineteenth century.

Rural America fought against the tide moving the country from a yeoman republic to an agro-industrial empire.[11] Much of this rural majority supported the Populist Movement[12] in its crusade against rail barons, bankers and land speculators. The Populist Movement, while a genuine threat to the capitalist classes, had run its course by the turn of the century, as it failed to unite farmers and farm workers with unions and factory workers in its battles against the urban establishment.

When the Great Depression hit America after the stock market crash of 1929, it was not only poverty itself that faced the country, but a major social contradiction. Farmers were losing their land because they had too much food to sell, while city dwellers went hungry because they did not have the money to buy food. America was a nation with 'breadlines knee-deep in wheat' (Poppendieck, 1986). Caught in oversupply predicaments, New Deal strategists working by market logic came up with what to them was an inescapable decision: plough the crops under before they saturated the market, as they did with one-quarter of the planted cotton acreage in the summer of 1933. To increase hog prices they slaughtered six million piglets, turning them into fertilizer tankage because the little pigs did not meet the packing houses minimum weight requirements. The public was amazed and angered by the Kafkaesque solution. Representative of the flood of letters to the Department of Agriculture was one from Chicago: 'I am writing to you to inquire if it is possible to get any of this pork, as I have been unemployed for many years and a lift along this line would be mighty welcome. I'll take a whole pig, dead or alive ...' (Poppendieck, 1986: 116).

Incongruous policies aside, New Deal agricultural programmes in soil conservation, credit relief, resettlement and mega-projects like the Tennessee Valley Authority were not without merit. Unfortunately, the benefits too often went to the large and relatively prosperous growers. After much promise and little performance, the Southern Tenant Farmers' Union claimed that under Franklin Roosevelt 'too often the progressive word has been the clothing for a conservative act. Too often he has talked like a cropper and acted like a planter' (Winthrop, 1982: 647). The legacy of New Deal's disproportional support for corporate farming has survived for over half a century. In the mid-1980s the largest 15 per cent of American farmers commanded 70 per cent of direct government farm payments. The system could be characterized as one that 'provided welfare for the rich and the market system for the poor'.[13] In a backhanded way, this 'market system for the poor' did evolve as a kind of solution to the farm problem. With Draconian ruthlessness, though obscured by market logic, it simply eliminated most farms and farmers. Without support, the bearers of the yeoman tradition formed invisible columns to start their migration to California and to the industrial heartland. Remarkably, 'emigration from the country to the city in the years since the Great Depression has been greater in numbers than entire immigration from foreign shores to the United States in the 100 years between 1820 and 1920' (Shover, 1976: xiv). The first majority was to become the last minority. In 1930 one in four Americans lived on a farm, in 1950 it was one in six, in 1970 one in 20, and then in 1985 just two in a 100 were left on the farm (Albrecht and Murdock, 1990).

The final devolution from yeoman agriculture to factory farming came in 1972 with the Russian wheat purchase.[14] In the largest agricultural transaction in the history of the world, the USA not only sold the USSR 19 million metric tons of grain, but lent the Soviets three-quarters of a billion dollars so they could make the purchase. The market circle was complete, from the slaughter of the little pigs to lending an arch Cold War enemy money to buy surplus wheat. Paradox, though, does not inhibit the man of action. Richard Nixon's Secretary of Agriculture, Earl Butz, called on farmers to respond to the new market opportunities by planting 'fence row to fence row', and advising them 'to get big or get out'. Dutifully, and in anticipation of the good times to come, farmers followed Secretary Butz's evangelical appeals. US grain production did rise 20 per cent from 1970 to 1981. Then came the inevitable and all too familiar collapse of saturated markets. Debt burdened, underdeveloped countries pumped out cash crop exports to earn money to pay off their long-standing loans.

Europe put up protectionist barriers, and the Russians could not buy enough grain to keep markets buoyant.

Through all of this, many of the farmers who did get big were just the ones who were getting out – being forced out. The grain boom had precipitated a land boom. An acre of Iowa farmland that, for example, sold for $419 in 1970 was worth $2,066 in 1980. Consequently, by just holding on to his land, a farmer's net worth quadrupled over the decade. With the prospect of surging markets, bankers encouraged farmers to borrow against their escalating equity in order to expand – buy more land and equipment to make even more money. But in the late 1970s interest rates shot up, just as prices were falling. Now, however, it was not the simple drop in income that pushed farmers to the wall. There were the mortgages against the land, the debts of expansion, that had to be paid back. Since land value fell right along with the grain prices, farmers were stripped of any net worth shield against their obligations. Their land was not worth enough to sell in parcels in order to survive to farm another day. Their options had run out. From 1981 to 1987, 26,000 Iowa farms, about 20 per cent of the total, went out of business. It was the Dirty Thirties all over again, but with so few farmers it was hard to get the rest of America to pay attention. Prosperous farmers bought their neighbours' land at thrift store prices, and, in some cases, insurance companies acquired land below market value as an investment, then hired bankrupted farmers to manage their properties, burning and bulldozing the farm homes, barns and silos to lower property taxes. In the process, the farms that did survive became even bigger, and today half of the food grown in America is produced on 4 per cent of the farms, with the top 5 per cent of the landholders owning 75 per cent of the country's private land.[15]

As the flow of grain moves back and forth across the planet, bought and sold in the form of futures contracts by traders who never plough, sow or harvest, it is not just individual farm families who absorb the shocks of the depressions in global commodity markets. Small-town America shares the pain, and the result is what Osha Gray Davidson calls 'broken heartland' (Davidson, 1990).

When farm income falls, a classic ripple effect is set in motion: main street businesses contract and close. In turn small-town and county tax bases shrink, and government revenues drop. Falling revenues lead directly to a decline in rural social services, just at a time when more families are in trouble and need the support of mental health agencies, counselling services, neighbourhood schools and local hospitals. Local banks fail along with the farmers they served. The merged banks do not necessarily equate their own interests with the survival of distant communities. Schools close and children spend the equivalent of one-third of the yearly instructional time riding buses to newly consolidated schools. Hospitals shut their doors to aging populations as Depression levels of *émigrés* leave rural America to seek their fortune, and their survival, in urban America.

Resilient and resourceful, those left behind in small-town America hang on. Trying to stay on the land or at least close to it while they wait for good times to return, farmers take full-time jobs in meat-packing plants, on construction crews or with the railroad, while their wives make beds at the town motel or take in 'industrial homework', sewing appliqués, for example, on designer sweat shirts, piece work that sometimes pays as little as $1.50 an hour. But the jobs, even the bad ones, do not always come. Consequently, homelessness is a problem not just of the inner cities. An increasing number of rural residents live under bridges, in cars, abandoned buildings and emergency shelters. Even when they are not driven to homelessness, more rural Americans are doubling up with friends and relatives. Finally, in the country's breadbasket, churches and community volunteers are setting up food banks to feed the hungry. Struggling to earn money to pay the bills, rural Americans cannot always find the time or the resources to grow their own food.

THE LATE TWENTIETH-CENTURY BACK-TO-THE-LAND MOVEMENT

In spite of the systematic underdevelopment of the American countryside, the rural and small-town way of life continues to capture popular imagination. As a case in point, just less than two decades ago US demographers identified what, for them, was a counter-intuitive population shift. They labelled this demographic anomaly the 'urban-to-rural migration turnaround'. The more than a century-long trend of relatively faster urban growth in America had started to reverse itself by the late 1960s and became in the 1970s a 'migration turnaround'. Through the 1970s, non-metropolitan counties, those not having a city with a population of over 50,000, had population growth rates exceeding metropolitan counties, those with urban centres of more than 50,000. Of course, only part of this counter-urbanization trend included back-to-the-landers, and much of it may be attributed to the low-density suburbanization of tracts of farmland and to upscale country residents who commute several hours a day to city jobs. But from the late 1960s through the early 1980s, the American neo-yeomen were part of a broad population movement that reaffirmed America's agrarian sentimentality (see Garkovitch, 1989).

The 1980s, however, brought a 'turnaround' of the turnaround migration. The farm crisis came between many Americans and their dreams of a life in the country. Metropolitan counties grew twice as fast in the 1980s as their non-metropolitan counterparts. The ripple effect of low farm prices, debt and foreclosure, a shrinking tax base and increased demands for social services pushed migratory flows from rural America back towards cities, reaffirming the century-old trend of rural decline. Depressed prices for country property in the wake of the farm crisis might have counter-balanced rural population losses by attracting urbanites looking to escape fast-paced city lives, but these potential urban expatriates needed the jobs that disappeared along with the rural population.

But the enduring attraction of small-town America continued to pull migrants from the country's metropolitan centres through the 1990s, in spite of often depressed rural economies. Non-metropolitan counties started growing again after the farm crisis of the 1980s, though, in contrast to the 1970s, their growth rates in the 1990s were lower than the growth rates for metropolitan counties. From 1990 to 1994, for example, non-metro counties grew by a total of 4 per cent. The same counties grew by only 2.3 per cent during the entire 1980s.[16]

The urban-to-rural migration turnaround, and the continued interest in country living through the 1990s, is a general phenomenon that parallels, but does not necessarily define, the back-to-the-land movement. But in terms of being able to specify the crystallization of a back-to-the-land movement or consciousness in American society, it is a publication rather than a demographic trend that serves as the specific historical marker. The periodical, *The Mother Earth News*, has been nearly coterminous with the late twentieth-century back-to-the-land movement itself. Founded just before Earth Day in 1970 by university drop-out John Shuttleworth, *The Mother Earth News* evolved from an obscure journal with a newsprint cover to a mass circulation magazine with a glossy exterior, filled with advertisements from corporate America. By the late 1970s and on into the early 1980s it counted on bimonthly press-runs of close to one million copies. For three decades it has served as a source of inspiration and information – some detractors would claim occasional misinformation – to millions of armchair homesteaders, and thousands of the practising variety, with articles like 'Grandma's Four Strand Braided Rug' and 'Mother's Bioshelter Greenhouse'.

Since Shuttleworth's *The Mother Earth News* did become a mass circulation magazine his views are not simply idiosyncratic, but evidently reflect the frustrations and dreams of hundreds of thousands of other Americans. The popularity of these ideas raises the question of the extent to which some of *The Mother Earth News* readers are able to translate back-to-the-land rhetoric to practice. In terms of a research problem, this question raises another one: How does the interested investigator go about finding neo-homesteaders in order to make an enquiry into their country experiences? The natural answer is to turn to the subscription lists of back-to-the-country magazines. And during the 1980s and early 1990s I used the subscriber lists of these magazines to conduct a series of interviews and questionnaire surveys of back-to-the-landers (see Jacob, 1997).

As my primary source for a sample of back-to-the-landers I selected the subscription lists of *Countryside* magazine of Withee, Wisconsin, since it is as much an information exchange, written by its subscribers, as it is a magazine in the traditional sense. At the time of my last survey, *Countryside* had a subscriber base of 30,000, though it currently fluctuates at around 50,000 subscribers. This survey, which serves as the basis for the profile which follows, drew on a US nationwide random sample of 200 *Countryside* subscribers for each of six states (California, Texas, Minnesota, Missouri, Maine and Georgia), with 1,200 potential respondents representing all the major demographic/census zones in the USA. The response rate for the survey was 58 per cent, and after the respondents not living on country property were eliminated from the analysis, there were 565 valid cases.

One of the first questions the *Countryside* survey can answer is whether back-to-the-landers are returning to their literal or mythic roots. To answer this question, the *Countryside* respondents were asked to report for themselves and their spouses how much farm experience they had before the age of 18, on a four-point scale from 'none' to 'a great deal'. Just over one in four (27 per cent) of the respondents, or their spouses for them, ticked the 'great deal of farm experience' category. For most smallholders, then, back-to-the-land appears more symbolic than literal. But there is another issue that needs to be raised in order to place this 'return' to the land in perspective. This related issue is the question of how much urban experience both the literal and metaphorical back-to-the-landers have accumulated during their adult years. The *Countryside* respondents, with an average age of 47, report an average of 15 years of living in an urban–suburban setting after the age of 18, and another 14 years of residence in the rural areas where they now live. Spending a decade or more working in the city to either save for homestead dreams, or temporarily explore urban options, appears to be a common route back to the country.

Coming back 'home' is typically something done in one's mature years after seeing the world beyond the edge of town.[17] Perhaps one reason smallholders are able to tolerate the provincialism of small towns is that they have the advantage of perspective, a perspective that, after years of city living, allows them to appropriate the virtues of rural life without overreacting to its deficiencies.

As one might expect, the idealism that prompts middle-aged, one-time urban residents to make a country move is closely associated with advanced levels of education. Almost three in five (58 per cent) of the *Countryside* sample have some university or academic college experience – with nearly one in eight (12 per cent) possessing a graduate degree, ranging from a Master of Arts in anthropology to a Doctor of Philosophy in wildlife biology. In addition to this academic course work, another one in five has one kind or another of post-secondary technical education, and less than one in ten (7 per cent) has failed to graduate from high school.

These well-educated urban expatriates, however, rarely accumulate sufficient financial resources during their urban sojourns to buy more than modest country property. Smallholder is an appropriate descriptor for the back-to-the-landers. The average land size for the *Countryside* respondents was 19 acres – a quarter of them have five acres or less and 75 per cent live on under 70 acres. And although their property sizes tend towards the modest, especially in the era of the mega-farm, the respondents do have a relatively secure tenure on their mini-homesteads. Over 90 per cent (92 per cent) of the sample either own the land they live on or are buying it – 42 per cent are mortgage free and 50 per cent are still making payments. One in 20 respondents rents, and the other 3 per cent of smallholders live with their families, on communal property or are part of a board and room arrangement.

This theme of modesty and stability carries over to the *Countryside* respondents' marital status. A substantial majority (80 per cent) of the survey respondents are married. Another 4 per cent report having common-law relationships, with the remaining 16 per cent divided among those never married, divorced or separated, and widowed. And for the more than eight out of ten from the sample who are married or living common-law, there is an average length of relationship of 15 years, for a group with an average age of 47. Large families in the eighteenth- and nineteenth-century farming tradition, however, have not resulted from these unions. On average the respondents have two children, with one child still at home.

Considering this information on the new pioneers' marital status, along with the other elements of this background profile, it is apparent

TABLE 8.1 Seven ways of living back to the land

Category	Description	(%)
Weekenders	Have full-time employment away from their farmsteads, but spend their free time (weekends, early mornings and evenings) working on their property.	44
Pensioners	Retired and supported by pensions (social security, investments and retirement plans).	18
Country romantics	Take part-time or seasonal work, then spend the rest of their time at work and at leisure on their property.	17
Country entrepreneurs	Major source of income comes from small business on property (cabinetmaking, welding) that does not directly involve farming.	15
Purists	Invest only part of their time growing a cash crop on their property, for just enough cash income to survive in a monetized economy; otherwise subsist from the resources of their own property and barter relationships with their neighbours.	3
Micro-farmers	Devote most of their working time to the intensive cultivation of cash crops on their property – usually fruits or vegetables with high market value.	2
Apprentices	Learn the back-to-the-land craft while working on someone else's farm.	1

that the back-to-the-landers are a group of people who have made substantial investments in their chosen way of life. They have committed themselves to long-term marriages; they have bought or are buying productive property; and they have invested time and energy in obtaining advanced education credentials. But the question now naturally arises as to just how these considerable investments can be translated into carving out an alternative economic space in order to earn a living in the country. In order to answer this question, I have constructed a typology of back-to-the-landers, which is built around two key elements of country survival: (1) the primary source of the smallholders' cash income (whether from or on their own properties or outside employment); and (2) the time they spend earning this cash income (working full- or part-time). In Table 8.1 there is a summary of the back-to-the-country types with corresponding percentages of respondents from the 1992 *Countryside* survey.

Table 8.1 does not support the stereotype of back-to-the-landers as carefree, urban dropouts who live at the margins of rural economies on rustic homesteads – the kind of smallholders that small-town residents call 'granolas.'[18] Only about one in seven (14 per cent) of the *Countryside* respondents, as 'country romantics', fit this category. At the same time more than two in five (44 per cent) are 'weekenders' who hold down full-time jobs. The relatively high number of weekenders in the sample suggests that trying to build a simple life in the country can become a rather complicated matter as one juggles full-time work to pay for the mortgage and farm equipment with the attempt at a relaxed appropriation of the country experience. The contradictions in the weekenders' approach to country living illustrate a dilemma shared by most neo-yeomen families. The dilemma comes in the form of an almost endemic tension between time and money. To the extent that back-to-the-landers take part-time work in order to experience the pleasures of subsistence farming, they are likely to find themselves without the financial resources to improve their properties or pay their mortgages. But if they try full-time work in order to accumulate the necessary finances to support their homesteading habits, they significantly reduce the time they want and need to work their land and, at the same time, enjoy the process.

One possible way to try to transcend the time–money dilemma is to work one's homestead as a 'micro-farmer', making a living from selling high-value produce in local markets. There are, however, inherent difficulties in both the current structure of the agricultural marketplace and the nature of the expectations that neo-yeomen bring with them back to the countryside that make micro-farming a less than attractive option.

Only 2 per cent of smallholders choose the micro-farmer way of living back-to-the-land (see Table 8.1). There are two basic reasons for this very low percentage of neo-homesteaders attempting a micro-farmer lifestyle. First, the agricultural marketplace has room for only a limited number of micro-farmers. The micro-farmers' potential customers are urban residents who are willing to pay higher prices for fresh, high-quality (often organic) produce, and undergo the inconvenience of shopping at distant farmers' markets or at the farm gate. Corporate mega-farms with their lower prices and worldwide supply networks (grapes from Chile in the middle of the winter) leave only niche markets for chronically undercapitalized micro-farmers. Secondly, successful micro-farmers have to spend long hours juggling a wide range of produce, from honey to free-range eggs to exotic herbs, in order to survive. This intense, market-driven way of life does not appeal to the neo-homesteaders anymore than it was attractive to their spiritual ancestors, the American yeomen. The 'pensioner', 'country romantic', country 'entrepreneur' and 'purist' back-to-the-land lifestyles are all built around the expectation of being able to apprehend mindfully the pleasures of a slower-paced country life (see Jacob, 1997). And most of the weekenders (44 per cent of the smallholders – Table 8.1) have the same set of goals, more often than not compromising the immediate enjoyment of homesteading while they attempt to accumulate the capital necessary for the long-term realization of country serenity.

What, then, are the prospects for a revival of the yeoman tradition? While present research indicates that 48 per cent of Americans would prefer to live in small-town and rural America (Brown et al., 1997: 416), with significant percentages of them presumably interested in a small-farm way of life, can one reasonably entertain the hope that more than a very small percentage could find alternative economic space and make their living from the land? Sanguine answers to these questions depend to a large extent on the public realization that factory farming violates fundamental market principles and in the process is systematically destroying the ecosystems on which it, and the rest of us, depend. The solution, though, to monopolistic factory farming is not so much government micro-management of global agribusiness as it is state-initiated social markets, in which neo-yeomen have the potential to demonstrate that corporate agriculture is not competitive when the complete range of the forces of production are taken into account.

CONCLUSIONS

Social markets can perhaps be best understood in contrast to nominally free markets. Over time the competitive dimensions of free markets engender monopolies and cartels as powerful corporations drive rivals from the marketplace.[19] The result consequently is anything but a free market, with a small and powerful number of corporate actors dominating commercial space, by a commonality of interests if not by explicit collusion. Social markets (Derber, 1995), then, using the countervailing power of the state, seek to re-institutionalize market principles. The primary method for re-balancing markets is not so much regulation as it is using market adjustments to force both producers and consumers to take into account all the factors of production, such as the environmental and social costs treated as 'externalities' by mainstream economists.

Contemporary agricultural markets provide a classic case for illustrating social market principles. The social market challenge to 'free-market' factory farming begins with the distinction between price and cost. It is certainly true that American consumers enjoy cheap food but, the environmental movement claims, it is also true that this low-priced food comes at a high cost. Costs associated with agribusiness production include the tax subsidies and support payments to growers, from which corporate farmers disproportionately benefit. Even more critical are the costs, not always easy to quantify, of ecosystem degradation that result from highly mechanized, petroleum-based farming: top soil loss, aquifer depletion and ground water contamination from pesticide, herbicide and fertilizer residues. This kind of chemical farming can take its toll in terms of rising healthcare costs for farmers and farmer workers, not to mention the potentially deleterious effect on consumers. Then moving from production to distribution, there are the transportation costs of highway construction and maintenance. But these economic expenditures do not take into account the social costs of farm foreclosures, the decline in small-town business and the migration of unemployed farmers to metropolitan destinations, all of which place an increasing burden on mental health, welfare and employment services.

In reaction to the diseconomies of scale that it sees in corporate farming, elements of the environmental movement, including the Wuppertal Institute and the World Resources Institute,[20] are calling for 'full-cost' pricing remedies. Businessman and social critic Paul Hawken has suggested the use of 'green taxes' in order to implement full-cost pricing (Hawken, 1993). Since the high social and ecological costs of factory farming are closely related to its use of large quantities of relatively inexpensive fossil fuel-based petro-chemicals, an environmental cost-recovery or green tax on these non-renewable resources would significantly increase costs for large-scale agriculture. Easily identifiable targets for green taxes include fuel (gasoline and diesel), pesticides, herbicides, fungicides and chemical fertilizers. Governments would then use the green tax revenues to undertake environmental clean-up campaigns. Agribusiness, of course, would have to pass the green taxes on to all consumers.

But since the back-to-the-land micro-farm operations are labour- rather than energy-intensive, they would be relatively unaffected by green taxes. As a consequence, the price of their produce could be considerably lower than that of the industrial farmers, and if the micro-farmer produce were sold in the region where it is grown, price increase to consumers would be moderate rather than dramatic. Ironically, then, back-to-the-landers as carriers of the yeoman tradition would move from obscurity to become major players in a full-cost, twenty-first-century agriculture. And this kind of change in the structure of American agriculture would give the neo-homesteaders the ability to transform their semi-subsistence avocations into full-time vocations. As a consequence, the rewards of smallholding would be as much financial as it is now emotional and recreational.

It is axiomatic, of course, that full-cost pricing would be vigorously resisted by agribusiness producers in danger of being priced out of the market, and by low- and middle-income consumers trying to keep their own food costs low. In anticipation of this kind of resistance, the advocates of full-cost pricing suggest a gradual phasing in of green taxes over a period of 20 years, and that the increased tax revenues from corporate farmers be used to reduce the tax burden on low- and middle-income earners. In addition, it is likely that consumers facing the immediate impact of higher-priced produce would attempt to lower their cost by organizing consumer co-operatives and buying directly from the growers at the farm gate or at farmers' markets.

However, this shift of the tax burden towards those who destroy ecosystems in the name of commerce is much more than a re-balancing of government revenues and expenditures in favour of the environmentally responsible. It also involves a radical reorganization of the production and distribution of farm produce that will require millions of Americans to move back to the countryside, into the alternative economic space created by the implementation of social markets. In a full-cost pricing scenario micro- and small farms would ring and intersect metropolitan territory, as well as fill in the empty spaces of suburban sprawl in cities like Houston, Texas,[21] as the supply lines between producers and their clients become radically shortened. And given the residual strain of agrarianism and preference for rural living in American culture, as represented by the back-to-the-land movement, the vacuum created by the exit of factory farmers should draw the necessary micro-farmers to repopulate the countryside and to feed their urban neighbours without the chronic surpluses that have impoverished generations of family farmers.[22]

While the implementation of full-cost pricing through green taxes is far from certain, particularly when considering the American political system's built-in tilt towards gridlock, its discussion in environmental and agricultural circles is a reminder that the labour-intensive smallholders, whether as yeomen or neo-yeomen, are still trying to escape their consignment to historical irrelevancy, to be periodically resurrected by curious scholars. Agrarianism itself in American culture has become progressively disconnected from agricultural production, as agribusiness interests have come to monopolize farming. But the back-to-the-landers throughout the twentieth century have kept alive the agrarian tradition in the alternative economic space of their semi-subsistent micro-farms. If the contradictions of factory farming continue to deepen and jeopardize agriculture itself, as environmentalists predict,[23] it is always possible that the back-to-the-land movement of the late twentieth century will be remembered not so much as an exercise of sentimentalism for a dying tradition as for the revival of agrarianism though its reconnection to agricultural production.

Notes

1 The mythological aspects of *The Wizard of Oz* receive comprehensive coverage in Nathanson (1991).

2 An overview of the demographic shifts affecting both rural and urban America are covered in Garkovich (1989). For recent reviews of urban-to-rural migration trends, see Johnson and Beale (1994) and Brown (1997).

3 The estimate of one million back-to-the-landers comes from Simmons (1979).

4 America's pre-eminent agrarian commentator is poet-professor Wendell Berry. His classic critique of American society is *The Unsettling of America, Culture and Agriculture* (Berry, 1977). Recent collections of Berry's essays can be found in Berry (1990, 1994, 1996). For an overview of agrarian thought which places Berry's work in context, see Montmarquet (1989).

5 There is an animated debate among historians over the character of yeoman life in the late eighteenth and early nineteenth centuries. The central issue of the debate turns on whether most early republican farmers were actively trying to sell their produce on large-scale commodity markets or whether they were primarily interested in subsistence agriculture and community trading networks. There is evidence to support both positions, depending on whether one draws on historical materials from the eastern seaboard, where farmers had ready access to markets, or from the relatively isolated interior, where they did not. Since there is a natural connection between yeomen who were apprehensive about the compromises to their way of life that commercial agriculture required and the contemporary back-to-the-landers who would like to be independent of the consumer culture, my interpretation follows that of Sellers (1991). Sellers contends that at least 50 per cent of the population in the early nineteenth century were subsistence or semi-subsistence farmers who resisted commodity markets and distrusted the commercial establishment. For a continuation and refinement of the debate, see the edited volume by Stokes and Conway (1996). For an overview of American agricultural history that is generally sympathetic to Sellar's point of view, see Danbom (1995).

6 'Enoughness' is a term used by Wolfgang Sachs (1989). In the same spirit of analysis, Atack and Bateman (1987) see the yeomen as 'Satisficers'.
7 Yeomen not only passively resisted the commercialization of farming, but at times they actively fought against what they saw as the oppression of courts and legislatures. From Shay's Rebellion in the 1780s to the farm resistance movements of the 1980s, there has been a long history of agrarian radicalism in America. Agrarian militancy has almost always prompted a reflexive counter-strike by the forces of established order. The Massachusetts Militia subdued Revolutionary War veteran Daniel Shay and his followers, and, as late as the Great Depression, rural violence in Iowa was sufficiently threatening to the governor that he declared martial law for several counties and mobilized the National Guard.
8 Agriculture has been, and continues to be, profoundly affected by the various transportation revolutions of the past 200 years, from canal and railways to container ships and refrigerated airliners. For an insightful history of the early stages of the transportation revolution, see Cronon (1991).
9 The conclusion of this paragraph follows the analysis of Henretta (1978).
10 This recounting of the events of post-Civil War America draws on Zinn (1980).
11 The phrase is adapted from Rischin (1976: x).
12 A standard academic reference for the Populist Movement is Goodwyn (1978). While Goodwyn is sympathetic to the populists and, by extension, to the yeomen, his sentiments are not universally shared by liberal historiography. A representative and influential source of this intellectual animus towards the populist and yeoman tradition is Richard Hofstadter, whom Eric Foner (1992: 597) called 'the finest historian of his generation'. Foner characterizes Hofstadter's perspective on nineteenth-century agrarian radicals in the following way: '[They were] small entrepreneurs standing against an inevitable tide of economic development. He [Hofstadter] saw them as taking refuge in a nostalgic agrarian myth or lashing out against imagined enemies from British bankers to Jews in a precursor to "modern authoritarian movements".' (Foner, 1992: 602). Christopher Lasch, who writes that Hofstadter's treatment of populist and yeoman themes is 'satirical and dismissive' (1991: 543), draws on Goodwyn's work in a vigorous effort to rehabilitate yeomen. By tracing the discontent with urban cosmopolitanism over the past two centuries, Lasch calls into question the assumptions on the unqualified desirability of material progress that analysts like Hofstadter appear to hold.
13 The 70 and 15 per cent figures are from Davidson (1990). The 'welfare for the rich and market system for the poor' quotation comes from Arthur M. Ford (1973) *Political Economics of Rural Poverty in*

the South, (Cambridge, MA: Ballinger, p. 92), cited in Davidson (1990: 29).

14 This discussion of the Russian wheat purchase draws material from Wessel with Hantman (1983).

15 The characterization of the farm crisis in this and the following paragraphs follows Davidson (1990). For a complementary UK perspective, see *The Ecologist* (2000).

16 The non-metro counties' growth rate of 4 per cent from 1990 to 1994 comes from an interview with Calvin Beale, senior demographer at the US Department of Agriculture, in *The Baltimore Sun Magazine*, 7 May 1995, p. 9. For a comparison of metro/non-metro growth rates in the 1980s, see Johnson and Beale (1994).

17 Dorothy in the *Wizard of Oz* is the archetypal back-to-the-lander.

18 Granola is a dry fruit, nut and cereal mixture favoured by backpackers and country nomads in general as a portable breakfast or snack.

19 For an analysis of how 'free' markets almost necessarily evolve into coercive ones, see Lux (1991).

20 The urls for the Wuppertal Institute and the World Resources Institute are www.wupperinst.org and www.wri.org respectively. These organizations carry up-to-date policy statements and papers on full-cost pricing and environmental taxes. In particular, see the World Resources Institute's conference proceedings, 'Environmental Policies in the New Millennium: Incentive-based Approaches', www.wri.org/incentives/index.html. See also, Chertow and Esty (1997).

21 In Houston, for example, Urban Harvest, a coalition of urban gardeners, has a membership of more than 80 community gardens spread throughout the metropolitan area.

22 The key question is not so much one of whether yeoman-style agriculture could feed America, or Europe, but whether the public, through the auspices of the state, is willing to give small and micro-farmers a fair return for their labour and their service in the interests of ecosystem survival. On the question of capacity it is instructive to note the following: 'And in the US, nearly one billion bushels of grain – half the nation's harvest – found no market in 1999, largely because the Asian economic slowdown reduced the demand for US farm exports' (Gorelick, 2000).

23 For a treatment of the environmental crisis that predicts we have only a few years to change radically our current levels of resource use and pollution in order to avoid ecosystem collapse and widespread starvation by the middle of the century, see Meadows, Meadows and Randers (1992).

References

Albrecht, D.E. and Murdock, S.H. (1990) *The Sociology of US Agriculture: An Ecological Perspective*. Ames, IO: Iowa State University Press.

Atack, J. and Bateman, F. (1987) *To Their Own Soil: Agriculture in the Antebellum North*. Ames, IO: Iowa State University Press.

Berry, W. (1977) *The Unsettling of America, Culture and Agriculture*. San Francisco: Sierra Club.

Berry, W (1990) *What Are People For?* San Francisco, CA: North Point Press.

Berry, W. (1994) *Sex, Economy, Freedom, and Community*. New York: Pantheon.

Berry, W. (1996) *Another Turn of the Crank*. Washington, DC: Counterpoint.

Bogue, A.G. (1963) *From Prairie to Cornbelt: Farming on the Illinois and Iowa Prairies in the Nineteenth Century*. Chicago, IL: The University of Chicago Press.

Brown, D.L. (1997) 'Continuities in size of place preferences in the United States, 1972–1992', *Rural Sociology*, 62: 408–28.

Chertow, M.R. and Esty, D.C. (eds) (1997) *Thinking Ecologically: The Next Generation of Environmental Policy*. New Haven, CT and London: Yale University Press.

Cronon, W. (1991) *Nature's Metropolis: Chicago and the Great West*. New York and London: W.W. Norton.

Danbom, D.B. (1995) *Born in the Country: A History of Rural America*. Baltimore, MD and London: Johns Hopkins University Press.

Davidson, O.G. (1990) *Broken Heartland: The Rise of America's Rural Ghetto*. New York: The Free Press.

Derber, C. (1995) *What's Left: Radical Politics in the Post Communist Era*. Amherst, MA: University of Massachusetts Press.

The Ecologist (2000) *Killing Fields: Its' Time to Tackle the Agricultural Crisis, The Ecologist* (Special Issue): 27–42.

Faragher, J.M. (1985) 'Open-country community, Sugar Creek, Illinois, 1820–1850', in S. Hahn and J. Prude (eds), *The Countryside in the Age of Capitalist Transformation: Essays in the Social History of Rural America*. Chapel Hill, NC: University of North Carolina Press.

Foner, E. (1992) 'The education of Richard Hofstadter', *The Nation*, 4, May p. 597.

Garkovich, L. (1989) *Population and Community in Rural America*. Westport, CT: Greenwood Press.

Goodwyn, L. (1978) *The Populist Movement: A Short History of the Agrarian Revolt in American History*. New York: Oxford University Press.

Gorelick, S. (2000) 'Facing the farm crisis', *The Ecologist*, 29 June, pp. 23–32.

Greider, W. (2000) 'The last farm crisis', *The Nation*, 20 November, pp. 11–17.

Hahn, S. (1985) 'The "unmaking" of the southern yeomanry: the transformation of the Georgia Upcountry, 1860–1890', in S. Hahn and J. Prude (eds), *The Countryside in the Age of Capitalist Transformation:*

Essays in the Social History of Rural America. Chapel Hill, NC: University of North Carolina Press.

Hawken, P. (1993) *The Ecology of Commerce: A Declaration of Sustainability*. New York: HarperBusiness.

Henretta, J.A. (1978) 'Families and farms: *mentalité* in pre-industrial America', *William and Mary Quarterly*, 25: 3–32.

Jacob, J. (1997) *New Pioneers: The Back-to-the-land Movement and the Search for a Sustainable Future*. University Park, PA: Pennsylvania State University Press.

Johnson, Kenneth M. and Beale, Calvin L. (1994) 'The recent revival of widespread population growth in nonmetropolitan areas of the United States', *Rural Sociology*, 59: 655–67.

Langley, N., Ryerson, F. and Woolf, E. (1989 [1939]) *The Wizard of Oz: The Screen Play*. New York: Delta.

Lasch, C. (1991) *The True and Only Heaven: Progress and Its Critics*. New York: W.W. Norton.

Lux, K. (1991) *Adam Smith's Mistake: How a Moral Philosopher Invented Economics and Ended Morality*. Boston, MA: Shambahla.

Meadows, D.H., Meadows, D.L. and Randers, J. (1992) *Beyond the Limits: Confronting Global Collapse, Envisioning a Sustainable Future*. Post Mills, VT: Chelsea Green.

Montmarquet, J.A. (1989) *The Idea of Agrarianism: From Hunter-Gatherer to Agrarian Radical in Western Culture*. Moscow, ID: University of Idaho Press.

Nathanson, P. (1991) *Over the Rainbow: The Wizard of Oz as a Secular Myth of America*. Albany. NY: State University of New York Press.

Poppendieck, J. (1986) *Breadlines Knee-Deep in Wheat: Food Assistance in the Great Depression*. New Brunswick, NJ: Rutgers University Press.

Rischin, M. (1976) 'Foreword', to John L. Shover, *First Majority – Last Minority: The Transforming of Rural Life in America*. Dekalb, IL: Northern Illinois University Press.

Sachs, W. (1989) 'The Virtue of Enoughness', *New Perspectives Quarterly*, 6: 16–19.

Sellers, C. (1991) *The Market Revolution: Jacksonian America 1815–1846*. New York: Oxford University Press.

Shi, D.E. (1985) *The Simple Life: Plain Living and High Thinking in American Culture*. New York: Oxford University Press.

Shover, J.L. (1976) *First Majority — Last Minority: The Transforming of Rural Life in America*. Dekalb, IL: Northern Illinois University Press.

Simmons, Terry A. (1979) 'But we must cultivate our garden: twentieth-century pioneering in rural British Columbia'. (Unpublished PhD dissertation, Social Geography Department, The University of Minnesota, MN).

Stokes, M. and Conway, S. (eds) (1996) *The Market Revolution in America: Social, Political and Religious Expressions, 1800–1880*. Charlottesville, VA and London: University Press of Virginia.

Wessel, J. with Hantman, M. (1983) *Trading the Future: Farm Exports and the Concentration of Economic Power in Our Food System*. San Francisco, CA: Institute for Food and Development Policy.

Winthrop, D. (1982) *The United States* (5th edn). Englewood Cliffs, NJ: Prentice Hall.

Zinn, H. (1980) *A People's History of the United States*. New York: Harper.

Conclusions: Re-making Geographies and the Construction of 'Spaces of Hope'

Roger Lee and Andrew Leyshon

Two of the ideas that underlie the exploration of alternative economic geographies in this book are the notions that (1) while all economies are irreducibly material, they are also social constructs; and that (2) the material and the social are mutually formative. This short concluding chapter attempts to draw out a number of the implications of this relationship by reflecting on some of the major themes which emerge from the substantive chapters which precede it. It does so by considering first, the diverse nature and practices of economic geographies and their inherent vulnerability, and then, the problematic practices of alternatives. In summary, the argument is that the simultaneously social and material nature of economic geographies ensures that all such geographies are vulnerable to breakdown and, therefore, that they are all not only utopian but 'alternative'. However, the possibility of alternatives does not necessarily lead to the construction of materially effective and socially widespread 'spaces of hope' as they may be diminished by material inadequacy, reform on mainstream principles or incorporation into the mainstream. Nevertheless, in demonstrating that economic heterogeneity is possible, the conception – and especially the practice – of alternatives do undermine the notions that history has come to an end and that we live in the singular best of all possible worlds. In short, whatever their material consequences and whatever their contradictory meaning in relationship to the mainstream, alternative economic geographies are a powerful reminder of the centrality of politics in economic life and of the continuing political optimism that underpins the sustenance of social life.

SOCIAL RELATIONS, MATERIAL PRACTICES AND THE VULNERABILITY OF ECONOMIC GEOGRAPHIES

Social relations of reproduction guide, shape and constrain material practices. If, then, the material is inescapable in the reproduction of social life, it may be practised in a wide social variety of ways. An implication that follows from this is that all economies are doubly fragile. They are vulnerable both to material breakdown or inadequacy, and to social resistance to their political acceptability. Material reproduction is, therefore, dependent on acquiescence in – if not support for – the social relations that define the way in which economic geographies function and develop. And this support is, of course, powerfully and recursively influenced by the material effectiveness of economic geographies. This is as true of prevailing and increasingly dominant neo-liberal forms of capitalism as it is of any other social form of economy. Hence, for example, the plethora of educational initiatives, personal financial advice and training which has the effects of tending to naturalize neo-liberal capitalism and to place it beyond critical question while reinforcing the belief that this is the best of all possible economic worlds.

However, to be materially effective all economic geographies (and not just those founded on neo-liberal capitalism) must be class-based in that they must be capable of producing, appropriating and redistributing surplus labour (Resnick and Wolff, 1987; see also Gibson-Graham, 1996; Gibson-Graham et al., 2002). In this sense, all economies are conflictual; all require extra-economic encouragement and belief to enable them to function. The social and the material come together in the class relation which demonstrates in practice their inherent inseparability and the double vulnerability of economic geographies.

UTOPIAN ECONOMIC GEOGRAPHIES

One consequence of the inseparability of, and mutually formative relations between, the social and the material is the continuing need to legitimize economic geographies in both material and social terms. In this context, a crucial element in sustaining the political acceptability of the social relations that guide material practices of economic geographies is the promise of a better (material and social) life that may be delivered through them (see for example, Harvey, 2000). The prospect (real or imagined) of improvements in material life and social well-being is a powerful means of eliciting support for the social relations that underpin economic geographies. In this sense, then, all economies/economic geographies are not merely doubly vulnerable, they are also utopian. How is this so?

The unavoidable exploitation implicit in all economies must be offset by the prospect at least of working for a better future. This, it may be argued, is made possible by access to the values created through economic activity. In some accounts it is also suggested that life enhancement may be induced even more directly by engagement in productive work. Thus the emergence of 'alternative' economies may reflect exclusion from, or dissatisfaction with, such processes. Such a response reflects the fact that the utopianism of economic geographies is inherent. Notwithstanding the distinctly less than utopian realities of economic practices and consequent appearances precisely to the contrary, economic geographies are constructed by and for people. Thus, despite – or because of – the widespread material and mental poverty and uneven power relations facing the majority of people as they engage in the process of making a living within neoliberal capitalist economic geographies, they must – if the reproduction of such geographies is to be sustained – be continuously persuaded that their involvement in them is the means of achieving this utopia.

ALTERNATIVE ECONOMIC GEOGRAPHIES: CONTRADICTIONS AND AMBIVALENCE

A consequence of the doubly vulnerable nature of economic geographies, and of the purposive characteristics of the relationships between people and the economic geographies through which they consume and produce the means of their reproduction, is that all economic geographies are not just inherently utopian but inherently 'alternative'. There is always an alternative, even if what we might call the ideology of convention, which tends to naturalize existing social and economic arrangements, makes it difficult to discern in day-to-day practice quite what that alternative might be. At the very least, then, all economic geographies involve a continuous process of material and ideological defence to sustain their own credibility. Faced with the competitive threat of 'other' economic geographies founded on different sets of social relations – and, no doubt, consequently involving different material practices – such defences must become even more focused. It is this potential of 'alternative' economic geographies to demonstrate a proliferation of possibilities, as much as their prospective material power, which endows them with such political significance. However, their relationship with their 'other' – the mainstream – remains unavoidable, ambivalent, unequal and full of contradiction. The preceding chapters have documented this complex relationship in detail and what follows is merely an attempt to highlight some of these contradictory relations.

Alternative economic geographies may be more or less radical. On the one hand, for example, Local Exchange Trading Schemes/Systems (LETS) are – potentially at least – highly radical. They presuppose nothing less than the displacement of established economic norms and practices with separate economies using autonomous means of regulation, evaluation and economic practice. Similarly, retro retailing also questions conventional forms of valuation and so demonstrates that one of the central relations of all economies, that of value, is also socially constructed – with all the transformative implications that such a finding has for mainstream norms of evaluation and work. Equally, the social economy and the continuing significance and dynamism of the informal economy point up, and so reinvigorate, the notion that communities of people construct economic geographies to enable the sustenance and development of their own lives and identities. On the other hand, credit unions bolster a financial system failing to deliver in crucial respects and so offer reformist support for the mainstream. And the same kind of comment could be made of employee ownership and the call for social markets.

However, things are not so simple. On the one hand, LETS may simply be incorporated into the mainstream as a way of offering cheap welfare services or commodities, retro retailing involves precarious, boring, lonely, self-exploitative and risky working practices reminiscent of self-employment, the social economy may serve simply to bolster a profoundly conservative and exclusionary notion of 'community', while informal work may be unevenly, even regressively, developed and offer less an alternative than a necessary but partial means of survival. On the other hand, credit unions invade the spaces of orthodox finance and empower those excluded from them, and the practice of employee ownership – whether judged 'successful' or 'unsuccessful' by mainstream norms – and of social markets offer not just a reminder of the active role of labour in economic practice and the experience of alternative social relations but demands the practical imagination to cope with the disjunction between them and the mainstream. Equally, alternative economic geographies are continuously open to incorporation by the mainstream and are faced with repeated resistance to the adoption of mainstream norms. Furthermore, they may merely supplement and/or ape the mainstream while remaining in a structural sense separate from it. Alternatively, they may offer support for the mainstream by providing material and social supplements yet not mount a challenge to its material or social power. Similarly, far from challenging the notion that economic geographies are constructed by the people who make them work, alternative economic geographies may be a means of reconstituting communities around more insistent economic norms. At the same time, alternative economic geographies may offer merely second-best solutions and so ghettoize their participants while intensifying exclusion from mainstream markets and repression of adequately resourced welfare provision.

Such contradictions and ambivalence are so significant that, in many ways, they serve to make the moniker 'alternative' almost meaningless except, as in some usages, as a term of opprobrium and marginality. However, the power of alternatives lies as much in the existence and the possibility of their existence as in their practice.

ALTERNATIVE ECONOMIC GEOGRAPHIES: IF NOT POSSIBLE, THEN NECESSARY

The unfolding tragedy that led to and followed from the events of 11 September 2001 led to a fundamental questioning, as well as a fundamentalist reassertion, of US hegemony and neo-liberal values.[1] The subsequent crisis of capitalism[2] in the wake of the accounting scandals revealed at Enron, Tyco and WorldCom, amidst the all too apparent fictive narrative of fictitious capital, may or may not have been proximately caused by the collapse of the World Trade Center but it exposed the *reductio ad absurdum* of the 'logic' of capital. This 'logic' suggests that even the holy grail of profit as a measurable and real category (which is the point and purpose of capital) can be sacrificed on the altar of accumulative competition. The vacuum of meaning that follows is, ironically, a crisis of *representation* that undermines the *material and social relations* through which capitalism is constructed.

But the response to this crisis is predictable – and not only in the elision (intended or unthought) of capitalism and 'market activity'. Consider the defence mounted by Samuel Brittan, the doyen of economic columnists, at *the Financial Times*. The problem, according to Brittan, is not the 'motivation [behind capitalism] but the framework of rules, conventions and assumptions within which market activity is carried out' (Brittan, 2002: 17). That such a response is made possible and justified is due to the prior assertion of 'the fact that competitive free market capitalism has been the biggest engine for growth – and for the reduction of poverty – that the world has ever seen' (ibid.).

This is undoubtedly true. But it is so only because (1) capitalism takes the credit for reducing the poverty that it was itself responsible for creating, (2) poverty is defined in exclusively financial and absolute terms along a linear trajectory which ignores the perpetual shattering of circuits of reproduction associated with uneven development, and (3) the apparency of the extra-*economic* inducement of increasingly intense global geo-*cultural* conflict is rarely questioned.

Nevertheless, the fact that it is still possible to argue in the manner of Samuel Brittan raises a range of questions both about why such arguments are possible and about the material and social power of capitalism and the (im)possibility of alternatives. The former are, in effect, the reason for the latter. That the possibility of alternatives is at best constrained leads to the conclusion that, precisely because alternatives are difficult to achieve and may not even be possible, they are ever more necessary.

Notes

1 See, for example, the remarkable spat between the World Bank and the IMF sparked off by the publication of Joseph Stiglitz's *Globalisation and Its Discontents* (2002).
2 According to Chris Chaitlow, an analyst at Collins Stewart, 'This looks like a major bear market of the sort you only (*sic*) get once every 30 or 40 years' (quoted in Elliott and Denny, 2002: 1).

References

Brittan, S. (2002) 'The rules need fixing, but greed can be good', *Financial Times*, 4 July.
Elliott, L. and Denny, C. (2002) 'Shares slump as panic hits stock markets', *The Guardian*, 4 July.
Gibson-Graham, J.K. (1996) *The End of Capitalism (As We Knew It): A Feminist Critique of Political Economy*. Oxford: Blackwell.
Gibson-Graham, J.K., Resnick, S. and Wolff, R. (eds) (2002) *Re/presenting Class: Essays in Postmodern Marxism*. Durham, NC: Duke University Press.
Harvey, D. (2000) *Spaces of Hope*. Edinburgh: Edinburgh University Press.
Resnick, S. and Wolff, R. (1987) *Knowledge and Class: A Marxian Critique of Political Economy*. Chicago, IL: University of Chicago Press.
Stiglitz, J. (2002) *Globalisation and Its Discontents*. London: Allen Lane.

Index

DATE DUE

JUN 1 1 2006			
DEC 0 4 REC'D			
OhioLINK			
MAY 0 5 REC'D			

GAYLORD PRINTED IN U.S.A.